Alex. Keith Johnston

# The school atlas of physical geography illustrated in a series

The elementary facts of geology, hydrography, meteorology and natural history

Alex. Keith Johnston

**The school atlas of physical geography illustrated in a series**
*The elementary facts of geology, hydrography, meteorology and natural history*

ISBN/EAN: 9783742846181

Manufactured in Europe, USA, Canada, Australia, Japa

Cover: Foto ©Klaus-Uwe Gerhardt /pixelio.de

Manufactured and distributed by brebook publishing software
(www.brebook.com)

Alex. Keith Johnston

**The school atlas of physical geography illustrated in a series**

# SCHOOL ATLAS

OF

# PHYSICAL GEOGRAPHY

ILLUSTRATING IN A SERIES OF ORIGINAL DESIGNS

THE ELEMENTARY FACTS OF

GEOLOGY, HYDROGRAPHY, METEOROLOGY
AND NATURAL HISTORY

BY

## ALEX. KEITH JOHNSTON

LL.D. F.R.S.E. F.R.G.S. F.G.S.

GEOGRAPHER IN ORDINARY TO HER MAJESTY FOR SCOTLAND; AUTHOR OF THE 'PHYSICAL
ATLAS,' THE 'ROYAL ATLAS,' ETC. ETC.

NEW AND ENLARGED EDITION

WILLIAM BLACKWOOD AND SONS
EDINBURGH AND LONDON
MDCCCLXIX

# PREFACE TO FIRST EDITION.

GEOGRAPHY has of late years assumed that position in the business of Education to which, from its interest and importance, it is so well entitled; and it is now found that, in order to do justice to its claims, it must be taught in a manner more systematic and orderly than was formerly considered necessary. The most obvious means of facilitating its study was the separation of its component parts into the great divisions of PHYSICAL or NATURAL, and GENERAL and DESCRIPTIVE GEOGRAPHY; and since public attention was first directed to the advantages of this distinction by the Author's folio Physical Atlas, many elementary treatises have been published for the purposes of general instruction. The subject-matter of each of these will find its appropriate illustration in the following Work, which has been prepared expressly for the purpose of conveying broad and comprehensive views of the Form and Structure of the Earth, and the principal phenomena affecting its outer crust.

Beginning with a representation of the Oceans, Lakes, Rivers, Mountain-Chains, Plains, and Valleys of the different portions of the Globe, it proceeds to the distribution of those elements by which its surface is affected—Earthquakes, Volcanoes, Heat, Wind, and Rain; and concludes with the actual occupation of its surface by the different races, families, and species of plants, animals, and man.

The object of PHYSICAL GEOGRAPHY, in treating of inorganic matter, is to represent the Earth in its natural state, divested of the accidental or artificial divisions which have been introduced by man's agency. By a patient study and careful comparison of numerous facts, the knowledge requisite for such a representation has, within a recent period, been greatly extended. Comparative Geography has led to analogous views of the structure of the different Continents similar to those which comparative anatomy has established in relation to the lower animals and man. The bones and arteries of the latter have their representatives in the mountain-chains and river-courses of the former. Asia, in its continental mass, presents a picture of majestic unity; Europe, indented and broken up into numerous peninsulas, exhibits an example of the greatest diversity; while the Western Continent is remarkable for its grand simplicity. Again, contrasting climate and position, we find that Africa, presenting the greater part of its surface to the burning rays of a tropical sun, has all its days and nights of nearly equal length—the diversity of the seasons being almost unknown; while, in the frozen regions around the Poles, night extends its absolute empire, "day disappears with its radiant cortege, or if it shines, it is but the longest meteor of a long night." *

Physical Geography teaches that there is an intimate and reciprocal action of Man on the Earth, and of the Earth on Man, without attention to which it is not possible to understand the national character or physical development of a people. In the East, his wants being easily supplied, and having no necessity to struggle with nature, Man resigns himself to indifference and fatalism; while in the West, in order that he may live, he must conquer the obstacles which nature opposes to his progress; and hence much of the energy, the perseverance, and the intellectual pre-eminence which characterise the races of the West. The whole character of a nation may, as Dr Arnold observes, be influenced by its geology and physical geography. "Who can wonder," he says, "that the rich and well-watered plain of the Po should be filled with flourishing cities, or that it should have been contended for so often by successive invaders?" † It is the abundance of its coal-mines that gives to England its pre-eminence in the manufacturing world. China is chiefly interesting to us for its cultivation of the tea-plant, the Molucca islands for their spices, and Siberia for its furs. It is owing to the nature of the soil and climate that the Southern States of North America are essentially agricultural; while, from having fewer advantages in this respect, and greater geographical facilities from seaports, the Northern States are almost as essentially manufacturing and commercial.

The eminently suggestive character of the Maps will, it is hoped, enable the intelligent teacher to draw many such contrasts and comparisons. Physical Geography is the history of Nature presented in its most attractive form, the exponent of the wonders which the Almighty Creator has scattered so profusely around us. Few subjects of general education are, therefore, so well fitted to expand and elevate the mind, or satisfy the curiosity of youth.

---

# PREFACE TO THE NEW EDITION.

IN the present edition four Plates appear for the first time,‡ the others have been revised and improved,§ and the text is in great part rewritten. For valuable aid in the design and execution of the geological illustrations, the Author is indebted to his friend, Archd. Geikie, Esq., F.R.S., Director of the Geological Survey of Scotland.

EDINBURGH, *December 1868.*

* RITTER.     † 'Lectures on Modern History.'
‡ Plates II. and III., Illustrations of Physical Geology; XIII. and XIV., River Systems of the British Isles
§ The Geological Map of the British Isles has been engraved from a new drawing.

# LIST OF PLATES.

# ATLAS

OF

# PHYSICAL GEOGRAPHY.

## DESCRIPTION OF THE PLATES.

### PLATE I.—Chartography and Climatography.

### CHARTOGRAPHY.

Fig. 1. The Mariner's Compass, distinguishing the true and the magnetic north points. The compass consists of a magnetised needle, bearing a circular card, the outer edge of which is divided into 32 points, half and quarter points, and into 360 degrees. The four chief or *cardinal* points are North, South, East, and West, the E. being towards the right when facing the N. The points in the middle space between two cardinal points are named after both, and written N.E., N.W., S.E., and S.W. A point half-way between one of these last and a cardinal point is named by a compound of the nearest cardinal point and the adjacent point: thus the point between N. and N.E. is termed N.N.E.; the point between E. and N.E. is E.N.E., &c. The points *next* the eight principal points have the word *by* between the name of such point and the next cardinal point: thus the point *next* to N. on the E. side is termed N. *by* E.; that next, on the W. side, N. *by* W., &c. The needle points to the *magnetic* north, which only in a few parts of the world agrees with the true or geographical north, and the difference between them is called the *variation* of the *compass*. The variation is termed easterly when the north point of the needle is drawn to the east of the true north, and westerly when drawn to the westward. The variation differs according to the place of observation on the surface of the earth, and the position of certain varying points of greater attraction called *magnetic poles*.

Fig. 2. A River-Course, with arrows pointing the direction in which the water flows. The *right* bank of a river is that on the right hand of a spectator situated with his back to its source, and the *left* bank that on the left hand.

Fig. 3. A Ship at Anchor. An *Anchorage* is a suitable holding-ground unimpeded with rocks, where ships can conveniently cast anchor. *Sounding* is the mode of ascertaining the

A

depth of water or the nature of the sea-bottom : it is performed by means of a lead attached to a sounding-line marked by certain divisions called *Knots*, at a distance of 50 feet from each other. Three knots make a marine league, and 60 knots are equal to a degree. A knot is another name for a nautical mile, so that when a ship is said to make 10 knots, she is progressing at the rate of 10 sea miles an hour. Soundings, or ascertained depths, are marked on sea-charts in fathoms. (See fig. 14.)

Fig. 4. A Lake or Inland Sea is a portion of water surrounded by land. *Woodlands*, or *Selvas*, land covered with natural timber. *Llanos* are vast plains of South America, alternately covered with rank vegetation and reduced to a desert state by drought.

Fig. 5. Firth or Estuary, an inlet of the sea connected with the mouth of a river. The term firth or frith is sometimes applied to an open channel, as the Pentland Firth, in which case it appears to have originated from the Latin word *fretum*; but as extensively used on the east coast of Scotland, the word is undoubtedly derived from the *fiord* (pronounced *fiorth*) of the opposite shores of Scandinavia.

Fig. 6. Gulf, a recess of the ocean or of a sea into the land. It is often used indifferently with Bay; but the latter is applied either to a large or small extent of water, while the former is used only to designate an extensive recess of the sea, or opening into the land.

Fig. 7. Bay, a large or limited recess of the ocean, of a sea, or large lake. A sheltered bay is a place of refuge for a ship in a storm; a *Harbour* or *Haven* is a small bay or port for shelter. (See *Gulf*, fig. 6.) The mouth or mouths of a river is a term used to designate that portion of country near where a river enters into a lake or estuary, or into the ocean. The French word *embouchure* is often so applied.

Fig. 8. Strait, a narrow passage of water between two continents or islands, or the entrance from the ocean to a gulf or lake. It is sometimes called a *channel* or a *sound*, and is often erroneously written " Straits." (See also fig. 6.)

Fig. 9. Delta, from the Greek Δ, a term applied to alluvial tracts between the bifurcating branches of a river. The best example of a delta is that at the mouth of the Nile, shown in the plate. An *Isthmus* is a narrow neck of land between two seas, joining two peninsulas or other portions of land. If the Isthmus of Suez, indicated on the plate, were removed, Africa would be an island. A *Cataract* is a sudden fall of a large body of water from a considerable height; the fall of a smaller body of water is called a *Cascade*. A *Rapid* is produced by a gradual declivity in the bed of a river. Rapids are sometimes navigable.

Fig. 10. Promontory, a more or less tapering projection of the land into an ocean, sea, or lake. *Cape* is a part of the coast extending into the water beyond the general shore-line, as in fig. 6; or the point of a promontory, as in fig. 11; it is also termed a *Point, Head, Ness* or *Nose*, and *Mull*. *Headland* is a high bluff portion of the coast not projecting far into the sea. *Archipelago*, a sea interspersed with many islands, as the Ægean Sea. *Breakers*, a name applied to rocks or banks which impede the motion of waves, and cause them to *break* or *foam*.

Fig. 11. Peninsula, a portion of land almost surrounded by water. When the neck of land which joins a peninsula to other land is narrow, it is called an *Isthmus*.

Fig. 12. Desert, a tract of sterile land, covered with sand or loose stones: the term is often applied to uncultivated wastes not devoid of vegetation. *Oasis* is an isolated spot in a desert, where perennial vegetation is supported by springs or by Artesian wells. *Marshes* occur in all situations—on heights, in valleys, in woods, or on the sea-shore; they are either of fresh or salt water. Some are constantly under water, while others are periodically dry. They often emit noxious exhalations termed malaria, which render the air unhealthy, and produce fevers.

Fig. 13. Chain or range of mountains, a series of elevations linked together continuously or closely, so that their length greatly exceeds their breadth: the *Crest* is the highest part of the range. *Peak* is the term applied to the conical or pointed summit of a mountain. *Table-land*, or *Plateau*, is a portion of the surface of the earth elevated above the general level of the region in which it is situated: it may be entirely surrounded by mountains or cliffs, or only bounded on one side by higher land. *Watershed*, or *Water-parting*, is the ridge-line, from the slopes of which waters flow in different directions. A watershed may either be elevated, as a mountain or range of mountains, or only a slight undulation above the general level. The space enclosed by a watershed is called a *River-basin*. A river-basin is the space, low or hollow with reference to the surrounding area, into which all the feeders of the main stream flow. A *Portage* is that portion of the watershed over which goods or boats may be transported from one river-basin to another.

Fig. 14. Hills or Mountains are represented on maps by means of shading, which is, or ought to be, darker in proportion as the height is greater. The shading is produced by lines, or *hachures*, placed alongside of each other, or by graduated tints. This scale shows the requisite depth of shadow necessary to represent any elevation, from a plain at the level of the sea to a slope of 45°.

*Roads*, or *Roadstead*, an open and exposed anchorage-ground, where ships are sheltered under lee of the land. *Sandbank* is an accumulation of sand under water, diminishing the depth of the sea. *Surf* is the name applied to waves which break upon a flat shore. The character of a coast-line, whether bold or flat, is shown on charts by signs indicating rocks, cliffs, dunes, sand, mud, shingle &c.

Fig. 15. The different forms of mountains, cliffs, &c., shown in profile. In the upper corner is a topographical plan of a *Glacier*, showing the descent of frozen snow, and a *Moraine* at the bottom.

Fig. 16 (upper section) shows the method of oblique hill-shading adopted in the new survey of Switzerland, and other Continental maps. The light is here represented as coming from one side only, thereby increasing the effect of relief. In vertical shading (lower section) the light is supposed to fall perpendicularly. According to this method—which is adopted in the trigonometrical surveys of England and France—the effect is less picturesque, but more accurate. Fig. 16 (lower section).—*Volcanic* peaks represented as they would be seen vertically, from a great elevation; and below, the same peaks as seen horizontally from the surface of the earth. (See Cone of a Volcano in Action, fig. 21, lower section.)

Fig. 17. *Contouring* is a method of representing, by lines on a flat surface, all those portions of a mountain or any high land which are of equal elevation. The shaded section is supposed to be a rock in the ocean, the bottom of which, *a h*, is on a level with the surface of the sea at low water: as the tide rises, the line of level, or *contour line*, is successively raised to 1 *g*, 2 *f*, &c. The lines under the section show the projection of the contours on a flat surface, according to the principles of surveying.

Fig. 18. Plan of an isolated mountain, seen vertically, with contours shaded according to the scale, fig. 14. The sections show the outlines of its several sides, with their elevations in feet, and the method of shading.

Fig. 19. A portion of Palestine, with the valley of the Dead Sea, to convey an idea of the method of representing absolute height and depth, vertically and in profile.

Fig. 20. Atolls, or lagoon islands, are circular reefs of coral formation rising out of the sea, and enclosing a lagoon. Other forms of coral reefs are shown in fig. 21.

## CLIMATOGRAPHY.

The names of the objects represented in the view of atmospheric phenomena are given by references at the bottom of the Plate.

*Atmospheric Refraction.*—When a stratum of the atmosphere, next the earth, is accidentally expanded or condensed, distant objects, instead of being elevated, are depressed; and occasionally, from two such strata of different densities being placed together, the objects are elevated by one of them and depressed by the other, and they appear double, one of the images being *direct* and the other *inverted.* The lower figure represents a ship with its masts and rigging: it is on the horizon, and is the actual vessel. Immediately above is an *inverted* image of the vessel, quite distinct in all its details. Owing to the great amount of refraction in the Arctic regions, the sun rises earlier above the horizon and disappears later than in lower latitudes, thus shortening, by several days, the dreary polar winters.

*Clouds* are vapours rendered visible by their condensation. They float in the atmosphere at an elevation varying from a quarter of a mile to nine or ten miles above the surface of the earth. Clouds are classified and divided into seven kinds, three of which, called the cirrus, cumulus, and stratus, are simple; and four—the cirro-cumulus, cirro-stratus, cumulo-stratus, and cumulo-cirro-stratus or nimbus—are compound.

1. The *Cirrus* (called by sailors the "cat's tail") is formed of thin parallel filaments, sometimes resembling a brush, at other times woolly hair or light network. It is the least dense of all clouds, and attains the greatest elevation and the greatest variety of form. The fine particles of which this cloud is composed are supposed to be minute crystals of ice or snow-flakes.

2. *Cumulus* is the name applied to convex or conical heaps of clouds extending upwards from a horizontal base. The structure of the cumulus is usually very dense; they are formed in the lower regions of the atmosphere, and are carried along the current nearest the earth. They often resemble bales of cotton, or distant hills covered with snow.

3. The *Stratus* appears in the form of a horizontal band, which generally forms at sunset and disappears at sunrise. It is a continuous sheet of cloud increasing from below upwards; and being the lowest kind of cloud, its lower surface usually rests on the earth.

4. The *Cirro-cumulus* is formed by the cirrus-cloud, by the breaking up and collapsing of its fibres into small rounded masses, separated by intervals of clear sky. It occurs in dry, warm, summer weather, when it is known as the "*mackerel sky.*"

5. The *Cirro-stratus* is a combination of the cirrus and stratus, consisting of slightly-inclined masses, dense in the middle and thin towards the edges. In form and position it often resembles shoals of fishes. It is a precursor of storms.

6. The *Cumulo-stratus* is formed by the blending of the cirro-stratus and the cumulus. It appears when the cumulus is surrounded by small fleecy clouds, just before rain begins to fall, or on the approach of a thunderstorm.

7. *Cumulo-cirro-stratus* or *Nimbus* is the rain-cloud or system of clouds from which rain is falling. It first assumes a black or bluish-black colour, but changes to grey when rain begins to fall.

A *Rainbow* is formed by the sun's rays falling on minute drops of water, as those of rain, or the spray from a fountain, which, on the opposite region of the sky, are refracted into prismatic colours, and reflected to the eye of a person placed with his back to the sun. Lunar rainbows are produced in the same manner by the light of the moon, but the bow, owing to the feeble illuminating power of the moon, is usually colourless.

*Lightning.*—If, by the sudden precipitation of the vapour of water in the atmosphere, a certain quantity of electricity is disengaged, then a spark is emitted, which passes from one cloud to another, or from a cloud to the earth, and the zigzag form of its passage is caused by the resistance of the air. The colour of lightning is generally a dazzling white, but it sometimes has a violet tinge. The noise of thunder, which follows lightning, is caused by the displacement of air by the explosion of the electricity, and the irruption of that which succeeds it filling up the vacuum.

*Waterspout.*—This remarkable meteorological phenomenon is mostly observed at sea, but sometimes on shore over sheets of fresh water. From a dense cloud a conical pillar descends tapering with the apex downwards. When over the sea, there are apparently two inverted cones—one projecting from the cloud, and the other from the water below it. Waterspouts originate in adjacent strata of air of different temperatures, running in opposite directions in the upper regions of the atmosphere, condensing the vapour and imparting to it a whirling motion. The water of the sea is not sucked up by a waterspout, as is commonly supposed, but only the spray that is carried up by the whirling vortex. The water poured out from waterspouts is always fresh, or nearly so.

The *Aurora Borealis* is a luminous electrical meteor of great beauty, seen in the northern part of the heavens, where the air is condensed by cold, and highly magnetic. It generally appears soon after sunset as a luminous arch, which spreads over the celestial hemisphere and speedily assumes a pyramidal form, with shooting columns of light on all sides, but mostly towards the zenith. Its forms subside and reappear with a flitting motion, and sometimes, after spreading over the heavens, the aurora bursts into a splendid display of variegated colours. The aurora which appears near the south pole is called *Aurora Australis.* The aurora is so extensive that it has been seen at the same time in Europe and America. It varies in height from 45 to 500 miles above the surface of the earth.

*Meteors.*—Igneous meteors, comprising fire-balls and *shooting-stars,* are luminous bodies which suddenly appear in the sky, generally at a great height above the earth, moving through the heavens with immense velocity, in a direction inclined to the horizon. After shining with great splendour for a few seconds, they sometimes explode with a loud noise. The shooting stars, which appear in vast numbers from the 9th to the 14th November, about the 11th August, and at other stated times, are supposed to form a group which revolves about the sun in an elliptical orbit, and which, in passing through the aphelion, come in contact with the atmosphere of the earth with a velocity of about 30 miles a second, when they become ignited and are consumed.

*Snow* is moisture in a frozen state which falls from the clouds when the temperature of the air containing aqueous vapour is at or below the freezing-point, 32° Fahr. If snow, in falling from the upper atmosphere, passes into strata of a higher temperature, it will be melted before it arrives at the earth's surface; hence it never snows in the torrid zone, nor in the temperate zone during the heat of summer, except on the tops and sides of very high mountains.

The *Barometer* is an instrument contrived to measure the height of a column of mercury supported by the pressure of the atmosphere. The scale attached to the glass tube on the left shows how much the mercury falls, for every 5000 feet, in ascending into the atmosphere : thus, supposing the temperature uniform, and the height of mercury in the tube to be 30 inches at the level of the sea, it would fall to 20½ inches at 10,000 feet, or to 9½ inches if carried to a height of 30,000 feet above that level. In measuring the height of mountains with the barometer, correction for difference of temperature between the base and summit, at the time of the observation, forms an important element in the calculation. The *Thermometer* is divided into degrees, according to the Fahrenheit, Centigrade, and Reaumur scales—the first of which is mostly used in this country, and the two latter on the Continent. The temperature at which water boils at different elevations depending on the pressure of the atmosphere, is used as a means of ascertaining the heights of mountains. Thus, as shown in the Plate, water boils at a temperature of 212° at the level of the sea, when the pressure is about 36 inches. If it were to boil at a temperature of 172°, it would indicate an elevation of 26,000 feet above that level.

In the scale on the right side of the Plate the different kinds of winds are named according to their estimated velocity in miles per hour. Thus, 50 miles per hour indicates a storm, and 80 miles a hurricane. The scale also shows the calculated amount of pressure, in pounds weight, exercised by each on a square foot of surface. Thus, a velocity of 50 miles per hour gives a pressure of 12,300 pounds on a square foot of surface. The scale on the extreme right of the Plate gives the ascertained amount of rain which falls in different places, showing the remarkable increase of quantity in proceeding from temperate to tropical countries: for while at London the annual fall is about 20 inches, in some parts of the Western Ghauts Mountains of India it amounts to 300 inches, or fifteen times as much.

### PLATE II. A.—Illustrations of the Action of Rain and Streams.

Over the surface of the whole globe there is everywhere in progress a process of disintegration and decay. From the tops of the highest mountains down to the shores of the sea, the rocks which come to the surface, even the hardest and most compact, are undergoing a gradual destruction. This is effected by a number of agents, chief among which are rain, streams, frost, glaciers, icebergs, and the sea. The result of the combined action of these different forces is the constant removal of solid material from off the surface of the land. The rocks decay, the mouldering parts are washed off, new surfaces are exposed to the same destruction, and give place in turn to others. This process of waste is known in geology by the name of *denudation*—that is, the denuding or laying bare of rock by the removal of other portions of rock which once covered it. In Plates II. and III. A. are given some illustrations of the chief forms under which denudation is carried on. The figures from 1 to 9a on Plate II. show the action of *rain* and *streams*, the left-hand column (1 to 6) representing the destructive effects, and the right-hand column (7 to 9a) the reproductive effects, of that action.

*Destructive Effects.*—Every year, in such a country as Britain, there falls a quantity of rain which, if collected together, would be sufficient to cover the whole surface to a depth of from 2 to 3 feet. Part of this rain sinks underground, and, after a subterranean circulation, reappears in the form of springs. Another and much smaller portion finds its way, by brooks, rivulets, and rivers, into the sea. When the moisture falls upon the land as rain it is nearly pure water, but when, after flowing over that surface, it comes down to the sea, it is loaded with mineral substances, either in mechanical suspension, such as fine mud, or in chemical solution, as is the case to a greater or less extent with all spring-water. These mineral ingredients or impurities are derived from the decay of the rocks of which the land is composed. When, therefore, we look at a muddy river entering the sea, we have before us a proof of the great waste which is going on over the surface of the region drained by that river.

In fig. 1 (Plate II.) the first and most direct effect of the fall of rain is seen in the print of the raindrop upon soft sand or mud. On a sandy or muddy beach the ground, after a shower of rain, is seen to be covered with thousands of little pits, produced by the impact of the raindrops. If a gale has been blowing at the time, the prints are slanted in the direction towards which the wind blew, and have the sand or mud ridged up on their farther side.

Fig. 2 illustrates the further progress of the same action. A thick mass of earth, with many stones of all sizes, is there shown to occupy the sides and bottom of a valley. The rain has gradually loosened and washed away the earth round each stone, but the latter has acted as a protection to the portion of earth lying immediately underneath it. Hence, as the soil around has been removed, the stones have appeared to rise out of the ground on pillars of earth. But the weather continues to act, though more slowly, upon these pillars. By degrees the support of the stone gives way, and then the column of earth, no longer protected, crumbles away. Instances of this kind on a small scale are occasionally to be seen in Britain, but they occur in great perfection in certain valleys of the Alps.

When the raindrops coalesce they form tiny rills; these unite into runnels, these into brooks, and so on till we come to the broad and stately river. Whether the stream be small or large, it makes a course for itself, which it tends constantly to widen and deepen. This it does by pushing mud, sand, gravel, or boulders along the sides and bottom of the channel. These materials rub down the rocks over which they are rolled, acting as a kind of file, and enabling the streams to cut for themselves a path through even the hardest rocks. In fig. 3 a river is seen to have in this manner excavated a long winding gorge or ravine.

In the carving-out of a river-course, however, much must necessarily depend upon the nature and arrangement of the rocks on which the water has to act. If they are very hard and compact, the rate of erosion may be extremely slow; if changes occur in their arrangement— softer portions, for example, alternating with others offering more resistance—the river may be turned from side to side in frequent curves. Sometimes a trifling obstacle is sufficient to turn a stream aside. A brook flowing along flat ground, for instance, may be deflected by a tree or a stone. Under similar circumstances a river winds in broad serpentine curves, as shown in fig. 5. It tends always to cut through the narrow part of the loops, and would thus shorten and straighten its course were it not at the same time busy making new loops elsewhere.

It often happens that a stream cuts out part of its course by means of waterfalls. This is specially the case with many ravines. A waterfall may be formed in the bed of a stream at the junction of two or more rocks which possess very unequal powers of resistance—such, for instance, as a hard limestone lying upon a soft shale. In fig. 4 a junction of this kind is shown by a section where we see the softer strata exposed to the dash of the spray from the fall, and crumbling therefore under the overlying harder strata, which project and give rise to the perpendicular waterfall. As the shales moulder, the support of the limestone from time to time gives way, and consequently large masses of rock are precipitated into the bed of the stream. But the soft beds underneath are again excavated, and a new outjutting ledge of limestone by degrees appears. In this manner slice after slice is cut off the face of rock behind the fall; the cataract recedes up the stream, and leaves behind it a deep gorge or ravine. In the figure (4), the end of a lake is seen to the left. When the ravine has cut its way back to the edge of this sheet of water, the waterfall will begin to diminish in height; and in proportion as it diminishes, the level of the lake will be lowered. In the end the waterfall will disappear, the lake, if not too deep, will be drained, and the river, flowing through the bed of the old lake, will escape through the narrow ravine which has been cut by the upward retrogression of the fall. The long deep gorge in which the river Niagara flows has been thus excavated by the backward movement of the famous falls.

That portion of rain which, instead of flowing off by brooks and rivers, sinks into the soil, sometimes produces striking changes upon the surface. Water percolating underground through loose porous strata removes portions of these, and tends to undermine the rocks that lie above. If this should happen on a slope, the overlying mass may be deprived of its support, and slide down or break off, forming what is known as a landslip. One of the most remarkable recorded landslips which have happened in this country is represented in fig. 6. It took place at Axmouth, near Lyme Regis, on the Dorsetshire coast. The cliffs there consist of chalk and greensand, resting upon stiff lias clay, which slopes towards the sea. There had been much rain during the early winter of 1839, and the rocks of the cliffs had become full of water, which, unable to descend beneath the clay at the bottom, flowed seawards along its surface underneath the cliffs. By this means the strata of the cliffs, which had probably been already undermined by the action of springs, lost their support. The ground cracked in many directions, and in particular a deep fissure was formed, three-quarters of a mile long, 100 to 150 feet deep, and 240 feet broad.

*Reproductive Efforts.*—The sediment worn off the land by rains and streams goes to form the material for the formation of new rocks. Some portions of it accumulate in valleys and lakes, but the greater part finds its way into the sea. Fig. 7 indicates the way in which a river sometimes deposits sand and gravel in a valley, and leaves it in the form of terraces on either side. Such river-terraces are of common occurrence. The upper ones are the oldest, for the rivers cut their way down, and leave their *alluvia* to indicate the former level of the watercourses.

Where a stream enters a lake, and the onward motion of its current is checked by the still water, the sediment which it held in suspension is allowed to settle to the bottom. As it gathers year after year, this sediment makes the lake at this part shallower, until a low flat tract of land is perhaps actually gained from the water. In fig. 8, a gain of this kind is indicated at the mouth of a mountain stream. In the figure underneath (fig. 8a), a section is given of the new ground thus formed in the lake. At the bottom lies the hard rock of the district in highly inclined beds. Over this come the sand and gravel pushed outwards by the stream. Each new addition of sediment, after travelling over the surface of these deposits, is shot gently over their further edge, which slopes away from the shore. Hence the alluvial materials are extended into the lake, very much as a railway embankment is made. They gradually rise towards the surface of the water, and at last, by the additions which they receive in floods, they get above the usual water-line. Vegetation then takes root, and helps to raise the general surface of the newly-formed ground.

The name given to such accumulations of sediment at the mouths of streams is *Delta*. Originally this term was applied to the great flat alluvial expanse at the mouth of the Nile (fig. 9), from the resemblance of its general outline to that of the Greek letter of the same name. But the word is now widely used to denote all such low alluvial grounds at the mouths of streams, whether large or small, and whether they have the delta shape or not.

The delta of the Nile (fig. 9) is one of the most symmetrical upon the globe. Its base next the Mediterranean is upwards of 200 miles long, and its length from the sea to Cairo 100 miles. Over this wide expanse of flat alluvial land the river sends innumerable branching channels, the two main arms reaching the Mediterranean at Rosetta and Damietta. All this area has been formed by the mud brought down by the river. It is prevented from extending seawards by a strong marine current which washes its base, and sweeps the sediment away into deeper parts of the Mediterranean.

In the case of the Mississippi delta, on the other hand, the sediment is allowed to settle in the Gulf of Mexico, and to push its way seawards in long branching arms. That arrangement is shown in fig. 9a, which represents the outer end of the delta. The mud and silt carried in suspension by that vast river are deposited in the Gulf, on either side of the narrow channels into which the main stream branches. These channels thus flow along narrow alluvial strips of land, which are pushed out into the Gulf at the rate of between 200 and 300 feet in a year.

The vast extent of the deltas of the Mississippi, Amazon, Ganges, and other large rivers, shows the enormous amount of *denudation* which has been carried on by the different agencies of atmospheric waste. These alluvial accumulations, hundreds of feet deep, and thousands of square miles in extent, have been formed out of the mud, sand, and gravel which have been worn away from the mountains and plains by the action of rains, streams, frosts, &c. Yet they do not by any means represent the whole amount of loss sustained by the land. A large proportion of the sediment is carried into the sea, and there mingles with the marine deposits, to form with them strata which, in after ages, will be elevated into hills and valleys like those from the rocks of which they have been derived.

**PLATE II. B.—Illustrations of the Geological Action of Ice and Snow.**

When water passes from the fluid into the solid state, as it does in the act of freezing, it assumes a new character as a geological agent, and begins a new series of geological changes not less remarkable than those which it effects when flowing as a liquid over the surface of the earth. Some of the more striking of these changes are illustrated in this Plate.

*Frost.*—All rocks are more or less traversed with joints, cracks, or other partings, into which water finds its way. When the water contained in these narrow chinks freezes, it expands, and in the act of expansion exerts a great pressure upon the walls of rock between which it is confined. After repeated freezings and thawings, these walls get gradually pushed farther and farther away from each other. Hence what was at first a mere crack becomes widened by degrees into a gaping rent, and the rock is split open. When this takes place along the face of a cliff, large masses of rock are from time to time detached, and fall to the base of the precipice. Fig. 10 shows a cliff of this kind traversed with perpendicular and transverse joints, of which frost has taken advantage to loosen and detach many large blocks. These are lying piled up at the bottom. There, however, they are exposed anew to the frosts and thaws of winter, and are still further split up. In climates where the temperature in winter sinks below the freezing-point, frost plays an important part in loosening the cohesion of soils and rocks.

*Snow.*—In mountainous countries where snow falls heavily, it sometimes accumulates in thick masses upon steep slopes. When springs or rain or the melted water from the snow runs down the slope underneath the snow, the hold which the latter has upon the ground is loosened; and its weight lending it impetus, a large mass of snow sometimes slides bodily down the declivity, sweeping trees, rocks, and houses before it, and carrying ruin far down among the vineyards or corn-fields below. This is known as an *avalanche* (fig. 11). The geological changes produced by the direct action of snow are, however, of slight moment when compared with those which are performed when the snow, solidifying into ice, moves down the valleys as massive resistless glaciers.

*Glaciers.*—In Alpine and Arctic regions there is a varying limit known as the *line of perpetual snow*, or more simply, the *snow-line*. The snow which falls below this line in winter is melted in summer, while the snow which falls on ground higher than this line is only partially dissolved by the sun's heat. If this latter portion, therefore, had no other way of disappearance, it would rapidly accumulate till the mountains and continents were deeply buried under snow.

This state of things, however, does not take place. The snow which gathers above the snow-line slides gently into the higher valleys, where, as it moves slowly downward, it becomes pressed into ice, and takes the form of *glaciers*. Where a broad mass of ground rises above the snow-line, the snow presses downward into every available valley, and the glaciers thus produced are the drainage of the snow-fields. A glacier is thus quite comparable to a river. It is indeed a *river of ice*, and carries off the superfluous snow somewhat as a river carries off the superabundant rain. In fig. 13 a sketch is given of part of one of the broad Norwegian snow-fields, and two of the valleys are shown into which the snow-drainage moves, each of them being in consequence occupied by a glacier. Fig. 18 is a map of part of the glacier district of Monte Rosa, and shows how the snow drains down into the valleys, and gradually takes the form of glaciers, which descend far below the limits of the snow-line. The general aspect of a glacier as seen from its lower end is given in fig. 12. On either side rise the mountainous sides of the valley ; in the centre, following all the windings of the valley, lies the glacier. In front is the termination of the glacier, where the ice melts away as fast as it is pushed forward, and where from the terminal cave a rushing rapid river takes its rise. Another glacier, which reaches the sea at Spitzbergen, is shown in fig. 14.

The direct geological results of glacier-action are twofold :—1. The carrying of enormous quantities of debris from higher to lower levels ; and, 2. The erosion of the sides and bottom of the valleys down which the ice moves.

1. The transporting power of a glacier is well seen in fig. 12. Along each side of the glacier large quantities of soil, stones, and blocks of rock, falling from the mountain-sides upon the ice, are borne onward, and are at last tumbled down in heaps where the ice finally melts. These are called *moraines*, those along the sides of the glacier being *lateral*, and those at its lower end *terminal moraines*. Where two glaciers, issuing from separate valleys, unite into one glacier, the right lateral moraine of the one and the left lateral moraine of the other come together, and form a central ridge or mound of rubbish down the middle of the enlarged glacier. This is known as a *medial moraine*. It is well seen in fig. 12. Upon the surface of the glacier shown in fig. 14 several lines of medial moraine are represented, indicating the union of a number of glaciers. Some moraines are of great size, and prove how great is the amount of waste going on in many mountainous districts.

2. The erosive power of glaciers gives them a high importance as geological agents. The sand, gravel, and stones which get entangled between the bottom of the ice and the rocky floor beneath it act as so many files to scrape, groove, and polish the rocks. Wherever the floor over which a glacier moves can be examined, it is found to present a smoothed and polished surface, which is covered with numerous straight and more or less parallel striæ, varying from the finest hair-lines up to grooves like the ruts of cart-wheels. This kind of surface is not, however, confined to the bottom of the valley, but can be seen along the sides where the ice has retreated a little ; and, indeed, often for many hundred feet above the present limits of the ice. The character of the striæ is indicated in fig. 19, which represents an ice-scratched stone from the moraine of an alpine glacier. Several sets of striæ are there indicated, the stone having several times shifted the position in which, jammed between the ice and the rock, it had its surface ground down and striated. The smoothed, polished, and striated surface given by the ice to the rocks on the sides and bottom of a valley is shown upon the rock in the foreground of fig. 15. This is known to geologists as a *roche moutonnée*. Two *perched blocks*, or large stones transported and left by the glacier, are also represented.

The extent to which a glacier is ceaselessly eroding the rocks over which it moves is strikingly proved by the muddy nature of all streams which issue from the lower end of glaciers. The mud is derived from the constant grinding away of the rocks within reach of the ice. The peculiar smoothed and striated surfaces of rock, only to be effected by glacier-action, can often be traced up mountain-sides many hundred feet above the present glaciers, and far down the valleys below the point where the glaciers now melt. Hence the existing glaciers are only the shrunk remnant of what they once were. It is now well ascertained that a large part of northern Europe has been in recent geological times covered with ice. Throughout the British Islands the ice-markings abound, not only in the valleys, but even on the sides and summits of the minor hills.

Where a glacier issuing from its own valley crosses another valley, it sometimes succeeds in damming back the drainage of the latter, and a lake is the consequence. Such has been the case in the valley depicted at fig. 15, where the two parallel lines sweeping along the sides of the hills mark two levels at which the water successively stood, when a barrier of ice stretched across the lower end of the valley, and prevented the water from escaping. They are lines of beach, the upper being the older of the two, and pointing to a long period during which the water, ponded back by the ice, stood at that level. A diminution of the glacier then lowered the level of the lake to the second line. Subsequently the glacier withdrew, the lake disappeared, and the drainage resumed its ancient course. These lines are well marked in some parts of the Scottish Highlands, where they are known as *parallel roads*.

*Icebergs.*—When, as in Arctic and Antarctic regions, a glacier reaches the sea, it breaks off there into large fragments, which float away and are known as icebergs. The seaward end of a Spitzbergen glacier is seen in fig. 14, and a number of fragments of ice are floating in front of it. In fig. 16 the aspect of an iceberg in mid-ocean is shown. The geological results effected by icebergs consist in the transport and deposit over the sea-bottom of such debris as may have fallen upon the surface of the glacier before it reached the sea, and also in tearing up and wearing away the surface of submarine rocks or mud over which they may be driven. Icebergs have sometimes been seen loaded with rock-rubbish, and sailing away into mid-ocean with this freight from Arctic lands.

*Coast Ice.*—In the far north the cold of winter suffices to freeze the water of the ocean. The first cake of ice formed along shore is borne up with the tide, and, augmented by renewed freezings, becomes at length a shelf of ice stretching for hundreds of miles along the coast, with a breadth of 120 or 130 feet, and a height varying from 25 or 30 to 60 or 70 feet. It is known in Greenland as the *ice foot*. The thaws of summer disengage millions of tons of rubbish, mud, earth, stones, and large blocks of rock, from the cliffs above. These are strewn upon the ice-foot. When the summer storms come, large portions of the ice-foot are detached, and float out to sea, carrying with them their piles of debris. In time they melt away, and their earthy burden is precipitated to the bottom. By this means the floor of the Arctic seas in many places receives annually large quantities of detritus, borne by coast-ice from the Greenland shores.

## PLATE III. A.—Illustrations of Sea-Action.

Among the various geological agencies by which the surface of the earth is continually modified, none presents itself under more varied and striking aspects than the sea. Covering, as it does, two-thirds of the entire surface of the globe, and washing thousands of miles of coast-line in every continent, it claims an important place alike in physical geography and in geology. Its waters are always in motion, and in moving they disturb the sand, gravel, or rocks over which they pass. On the calmest summer-day, when the surface of the ocean is scarcely dimpled by a gentle undulation, we may yet watch the wavelets, as they ripple over the beach, catch up the fine sand, and roll it backwards and forwards as they advance and retire. The particles of sand, in rubbing against each other, are necessarily rounded and lessened. Now from this weak and hardly noticeable action we may trace a gradually increasing exhibition of force, until we behold the billows of a winter storm driving huge blocks of rock shorewards, and bursting in clouds of foam upon the cliffs.

Whether it be gently or with violence, there is a ceaseless tear and wear in progress all round the margin of the sea. At considerable depths beneath the surface of the ocean there is probably little or no action of this kind; but where waves, tides, and currents can act, as they do when they come against a coast-line, the land suffers a gradual loss. In this process of destruction it is not merely that the sea itself beats against the land, and wears it down. Frosts, springs, rain, acting upon the rocks of the shore, corrode and loosen them, and in this way powerfully aid the waves. In many cases, indeed, the waves do little more than remove the detritus which has been produced by these atmospheric agents of degradation. Much also depends upon the set of the tides, the force and direction of the prevalent winds, as well as on the form of the coast-line, and the nature and arrangement of the rocks against which the sea breaks.

Plate III. A. contains some illustrations of different aspects of sea-action. In fig. 1 the rocks dip down beneath the waves like a kind of natural breakwater. In such a case, it is evident that the breakers must burst upon the slope and roll up towards its summit without doing it much damage. Their action is here reduced to a minimum. In the end, however, they succeed, with the help perhaps of frosts and atmospheric disintegration, in loosening some part of the rocky declivity. Other motions follow, until the slope is planed down.

Fig. 2 shows how the waves advance when, in place of a slope, they have obtained a vertical face of rock to undermine. The pieces of stone detached from the cliff by frost or otherwise fall upon the beach, and are there made use of by the waves. In a gale these stones are lifted by the breakers, and swung forwards against the base of the cliffs from which they have fallen. Year after year, as this grinding goes on, the stones, originally rough and angular, get rounded and polished, until they sometimes are not unlike cannon-balls. And indeed they are really missiles, while the breakers which drive them on form a kind of sea-artillery. In many places they have battered the base of a precipice until they have hollowed out a cave, into which the waves rush with their freight of gravel and boulders. In other cases they have so undermined the cliff (as shown in fig. 2), that at last the overhanging portion fails to have sufficient support, and topples down upon the beach.

The next two figures illustrate further stages of this action. In fig. 3 a sandstone cliff full of points has been tunnelled, partly by the waves and partly by rain and frosts. The roof of this passage will crumble away while the sides are being demolished, until in the end the arch falls in, and a detached mass of rock is left in advance of the cliff. This has already been the case with the sandstone pillar on the left hand. Formerly it was a part of the main cliff, but by degrees it has been severed, and year by year its bulk is diminishing, under the combined influences of the atmosphere and the waves. In fig. 4 are shown different stages, from the arched passage to the mere sunken reef or skerry. First a hollow or cave is made upon an out-jutting mass of cliff; the cave is continued till it pierces the headland, and a tunnel is produced; this is widened and heightened, and at length the roof falls in. A detached sea-stack now rises up in front of the cliff; slowly it loses bulk as centuries pass away, until at last it is reduced to but a tangle-covered rock, over which the sea-swell is ever breaking. Such has been the history of the newly-submerged rocks on the right of the figure. Farther to the left rises a pillar of rock now detached from the headland; then comes the arch which is now being worn away, and lastly the main line of cliff, out of which future sea-stacks and inlets remain to be carved as the sea advances inland.

When the waves beat upon hard rock, their rate of progress is necessarily very slow. But when they encounter only sand, clay, or other soft materials, they sometimes march onward with a melancholy rapidity. Figs. 5 and 6 represent a remarkable example of such a rapid inroad. In the former figure the old church of Reculver, on the Kentish coast, is represented as it stood in the year 1781, separated from the sea by a considerable space of ground covered with the churchyard and some buildings. Fig. 6 (Lyell) is from a drawing made in 1834. The intervening ground has all been swept away, human bones have been seen protruding from the cliff under the churchyard, and were it not for the artificial breakwater which has been constructed, the church would probably by this time have wholly disappeared. Similar evidence of the advance of the North Sea is abundant along the eastern coast-line of England. In some places the cliffs consist of soft stratified materials (as in fig. 7). In other cases the cliffs are formed of stiff unstratified boulder-clay (fig. 8). Where, as in the latter case, blocks of hard stone or boulders occur abundantly in the substance of the cliff, these accumulate below, and form a kind of temporary breakwater. Sometimes, and more especially where a river enters the sea, a bank of gravel and sand is thrown up by the waves in front of the land, from which it is separated by a channel of still water (fig. 9). In such instances the waves raise up a barrier against their own advance, and protect land which otherwise could not fail to be destroyed. In reality, however, the materials of these natural breakwaters do not withstand the waves. They travel along the coast-line even for many miles, driven by the prevalent direction of the breakers; and at last, where the bank ends, are either driven ashore or swept away seawards.

The general result of the destructive action of the sea is to pare off and level down the margin of the land. As it is only the parts of the sea near the surface—those exposed to disturbance by winds—which display this action to any considerable extent, there is a constant tendency towards the production of a plane, the surface of which shall lie beneath the limit to which the influence of wind-waves extends. This surface is known to geologists by the name of a *plane of marine denudation.* We often see it displayed on a small scale along our rocky shores, as in the diagram (fig. 10), where a series of inclined strata dipping inland has been planed down, the worn edges of the beds forming a kind of level surface under the waves. The result has been achieved, however, on a vast scale in some countries, broken and crumpled rocks having been, under the combined forces of denudation, planed down into wide plateaux or table-lands. Traces of such forms of surface are still to be detected in this country. In Wales, for example, and among the Scottish mountains, geologists have found relics of ancient table-lands, out of which the present system of valleys and lakes has been excavated.

The amount of solid material which is worn off the land every year by the united efforts of rain, frost, springs, glaciers, rivers, and the sea, cannot but be enormous. This material, however, is not lost: carried out beneath the sea, it is there spread over the bottom in layers of sand, silt, mud, and gravel. Year after year these layers increase in depth and solidity. In after ages, when the bed of the sea comes to be elevated into dry land, these deposits will, to a large extent, have become by that time hard rock, and will form then the hills and valleys of the new land. Such has been the origin of most of the hills and mountains now visible upon the surface of the globe. They consist in great part of sandstones, limestones, or other strata, which once were deposited as sediment over the floor of the ocean, but which, in the course of many revolutions of the earth's surface, have now come to form hard parts of high and rugged land.

The Atlantic is the only one of the oceans whose bed has been at all explored, and our knowledge of it is still limited to that part of it which lies north of the equator. In no direction does this portion of the ocean bed present the form of a smooth basin from shore to shore, but, on the contrary, a series of great valleys and table-lands, with slopes as steep as any to be found on the land of the globe. This will be evident from a glance at the sections across the basin of the Atlantic (fig. 11), which have been chosen as nearly as possible at right angles to one another; but in looking at them it must be borne in mind that the vertical scale of each has been exaggerated to twenty-five times the reality, without which enlargement the undulations would scarcely be visible. The section (a) from Trinity Bay, Newfoundland, to Valencia Island, Ireland, on the line of the telegraph cables, crosses one of the shallower regions of the North Atlantic, and shows probably the most generally level part of the whole ocean bed. The greatest depth in this section is 2400 fathoms, a sinking which corresponds to the elevation of the highest summits of the Alps. The section (b) from Cape Race, Newfoundland, to Cape St Roque, Brazil, shows one of the most uneven parts of the ocean-bed. It passes first over the bank of Newfoundland, which stretches south-east from the island for more than 300 miles, at a depth of less than 100 fathoms, plunges suddenly into the deepest part of the ocean, first to 4000, then to 6600 fathoms, perhaps the greatest depth in the North Atlantic—a depth in which the highest mountain of the Himalayas might be set with its peak still 10,000 feet below the level of the sea; rises again by a steep slope to the plateau which fills the centre of the North Atlantic, culminating in the Azores Islands, and sinks once more before reaching Cape St Roque, in the depths which lie in the narrow part of the ocean between the West African and South American coasts. The section (c) from Cape St Roque to Cape Palmas, West Africa, crosses a great ridge in mid-ocean, which possibly extends from St Helena to Ascension and St Paul, having these islands for its summits, and thus dividing the ocean-bed here into two distinct basins.

### PLATES III. B & V.—Earthquakes, Volcanoes, and Rising and Sinking of Continents.

### EARTHQUAKES.

Earthquakes and Volcanoes are phenomena intimately connected with a heated condition of the interior of the earth. The concussions of earthquakes are communicated from below, and consist in commotions of parts of the earth's surface, sometimes so slight as to be almost imperceptible, at other times producing violent convulsions, which leave lasting memorials of their destructive effects. Earthquakes generally advance in a linear direction, undulating, with a velocity of from 25 to 35 miles in a minute; but sometimes in circles or great ellipses, in which, as from a centre, the vibrations extend with decreasing force towards the circumference. Their duration is often extremely short, the most destructive shocks being over in a moment. The great earthquake of Lisbon in 1755 lasted only six minutes.

Earthquakes are generally accompanied by detonations, the noise of which has been compared to the roll of thunder, the rattling of heavy waggons, the irregular discharge of cannon, or the hollow sound of an exploding mine: sometimes these sounds are heard for a considerable length of time without any shock following; and, on the other hand, cases occur in which the severest shocks are not accompanied by any noise whatever. When the earthquake wave, or that undulating motion of the earth's crust producing the phenomena of the earthquake, extends under the bed of the sea, the latter is always agitated. During the earthquake at Lisbon the sea rose 50 feet above its ordinary level. In the open ocean the effect of an earthquake upon a ship resembles that produced by striking on a rock under water.

An illustration of the effects of an earthquake upon the surface of the ground is given in Plate III. B, fig. 5. Towards the left a rounded building has been shattered, and one portion has sunk down to a much lower level than the remaining part. On the right the ground is represented as having opened in a long ragged fissure. Into such rents houses, trees, human beings, and different animals are sometimes precipitated; and if the rent close again immediately, all trace of the objects engulfed is lost. It may happen, however, that a second shock coming close after the first may reopen the fissure, and even eject the buried objects to the surface again.

DISTRIBUTION OF EARTHQUAKES.—In so far as yet known, these may occur at any part of the earth's surface. No country, whatever the nature of its geological formation, is entirely exempt from such visitations: they have occurred even in the loose alluvial soil of Holland. They appear with equal power in all zones and at all seasons; and it is probable that the surface of the globe is continually agitated by concussions in some of its points.

The earthquake district of the *Mediterranean* and of *Central Asia* extends from the Azores and Canaries on the west to Lake Baikal on the east, and probably to the Great Desert of Africa, the Nile Delta, and the Desert of Arabia on the south. It is the longest and most regular zone of volcanic action on the globe.

From the earliest times volcanic appearances have been observed on a line extending between the Caspian Sea and the Azores, which is the centre whence the earthquake proceeds, and according to distance from which the shocks decrease in frequency and violence; and it appears that the boundaries of the principal concussions are limited by the more or less parallel position of the Pyrenees, the Alps, Carpathian and Caucasian mountain-chains, in reference to this central line.

Within this district the *Earthquake of Lisbon* occurred on the 1st November 1755. Suddenly a sound of subterranean thunder was heard, which was immediately succeeded by a violent shock, demolishing the greater part of the city, and within six minutes 60,000 persons perished. Its movement was undulatory, and it is estimated that it travelled at the rate of 20 miles in a minute. The axis of the shock formed a line extending from the coast of Marocco, along the western shore of Portugal, to Cork in Ireland. From Lisbon as a centre the line of devastation extended north to Oporto and south to Ayamonte. Within this space the sea, fearfully agitated by the concussion of its bed, caused great destruction. At Cadiz a wave 60 feet high rolled over the land; at Lisbon the sea rose 50 feet; at Funchal in Madeira it rose 15 feet above the highest water-mark; and at Kinsale in Ireland it overflowed the marketplace. The space within which this earthquake was observed on land, extended in an elliptical form from the island of Madeira to Abo in Finland, and from the north of Scotland to Sardinia. In Scotland the water in Lochs Lomond, Katrine, Long, and Ness rose and fell repeatedly to an extent of 2 or 3 feet, and a similar movement of the surface occurred in the lakes of Germany, Switzerland, France, and Scandinavia. The agitation of the waves of the sea extended across the Atlantic to the Antilles, where the islands of Antigua, Barbadoes, Guadeloupe, and Martinique were overflowed; and at the same time an unusual movement was observed on the surface of Lake Ontario.

Italy and its islands have been more frequently visited by earthquakes than any other part of Europe. Pompeii and Herculaneum were thus destroyed in A.D. 63, and sixteen years afterwards, the cities, which had in the mean time been rebuilt, were buried under the ashes and lava ejected from Mount Vesuvius. Since that time earthquakes have been of frequent occurrence in Lower Italy and Sicily, especially in the eighteenth and nineteenth centuries. The great earthquake of *Calabria*, 5th February 1783, is remarkable for the concentration of its force, since within an area of 500 miles 300 towns and villages were destroyed, nearly 100,000 of the inhabitants perished, and the face of the country was completely changed. The convulsions extended along the granite range of the Apennines from north to south, but were not perceived to the eastward of that chain.

North of the foregoing district, in Basilicata and Principato Citra, with the Val di Diano as a centre, a series of earthquakes commenced on the 16th December 1857, and continued, with intervals, during January 1858. These exhibited many of the most appalling features of the phenomena; hills were levelled, and deep fissures formed. Upwards of a hundred hamlets, villages, and towns were injured, many of them ruined, and nearly 10,000 persons perished.

In 1838 a region of Southern Europe, far to the eastward, was visited by a destructive earthquake, the shock of which was felt at Vienna and Constantinople, in Southern Russia, and the countries of the Lower Danube. Greece and Turkey have also been visited by frequent and very violent earthquakes. Hence this district extends over Asia Minor, Syria, and Palestine, to the Caspian Sea and the Caucasian Mountains, where the country in the vicinity of Teflis has suffered severely, both in earlier and later times. Persia and the countries of the Caspian form the connecting link between this and the *district of Central Asia*, where the course of linear concussion follows the direction of the great mountain-system of Thian Shan or Syan Shan, north of which the eastern portion of the Altai Mountains, the Plutonic fissure of Lake Baikal, and the warm springs of the Orkhon River, form the central line of that volcanic convulsion which destroyed Karakhorum at the end of the thirteenth century. The countries of Asia have suffered as much from earthquakes as those of Europe. The great earthquake of Syria, in May, A.D. 526, most severely felt at Antioch, is said to have caused the sacrifice of 250,000 lives; and in the same region, in 1837, a shock occurred over an extent of 500 miles in length by 90 in breadth, by which 6000 persons perished. The earthquake of Cutch, in 1819, caused such alterations that the bed of the eastern branch of the Indus, previously only 1 foot deep, was suddenly depressed to a depth of 10 feet. In 1832 a shock passed through the chain of the Hindoo-Koosh, and extended to Bokhara and Kokand.

*Earthquake District of India.*—Between the years 1800 and 1842 one hundred and sixty-two earthquakes are recorded. In 1843 twenty-three shocks were observed, and since then four or five have been recorded in India each year. Most of these occurred within the region of the Himalayas, including Cabool, Jellalabad, Cashmere, Nepaul, and Assam. They very seldom happen in the peninsula south of latitude 15°. Some of these were very severe. In 1833 a shock occurred in the plains between Delhi and Chittagong, within the elliptical space marked in the Map, the undulatory movement of which was from north-west to south-east: in the valley of Nepaul 4000 houses were overturned, 414 persons were killed, and 172 wounded. In November 1842, Calcutta was nearly the centre of an emanation, the shock of which extended to Darjeeling on the north, Chittagong on the east, and Monghyr on the west. From Mount Aboo as a centre, a shock extended in an easterly and westerly direction, from Surat in the south to Simla in the north, over an area of about 100,000 square miles.

The *Earthquake District of Iceland* probably includes the whole of Great Britain, the north of France, Denmark, and Scandinavia; it extends to Greenland on the north and west. Its connection with the Mediterranean district is shown by the coincidence as to time in the great eruption of Skaptar Jokul of 1783 with the earthquake of Calabria in the same year. The first-recorded earthquake in England occurred in A.D. 974; since that time slight shocks have been felt in all parts of the British Islands, and latterly they have been very frequent at Comrie, in Perthshire.

In *Africa*, earthquakes appear to be unknown, except in a part of Barbary in the north, the region of the Red Sea in the east, and the Cape Colony in the south.  In the latter region only three shocks are recorded in the present century, all of them slight—one in 1809, another in 1835, and the third in August 1846, which latter extended over an area of about 400 square miles.

Earthquakes are known to have occurred in almost every other explored region of the Old World, but they are most prevalent in the volcanic girdle marked by shading on the Map.

*Earthquakes of America.*—Both as regards frequency and extent, these equal those of the Old World.  They occur chiefly along the coasts on the western side of the Andes, and on the northern declivity of the mountains of Venezuela, very few shocks having been recorded as taking place to the eastward of the Andes.  Among the most remarkable was the *earthquake of Chile*, 20th February 1835, which was felt everywhere between the Island of Chiloe in the south and Copiapo in the north.  All the towns and villages between the parallels of 35° and 38° south were destroyed, and the line of coast was permanently raised.  In the great *earthquake of Caracas* in 1812, the concussion, like that of Lisbon, formed an ellipse.  It was felt from near Bogota in the west to the Gulf of Paria in the east, and formed part of the district of nearly synchronous earthquakes which occurred between 1811 and 1813.  One of the great phenomena of tropical America was the eruption of *Coriguina*, and the synchronous earthquake of New Granada in 1835.  A few weeks after the eruption of this volcano, nearly the whole of New Granada was violently convulsed.  The *earthquake of Guadaloupe*, 8th February 1843, is one of the most interesting on record.  It was felt over a long narrow zone, from New York to the mouth of the Amazon, 3000 miles; but its most violent concussion was confined to Guadeloupe and Antigua.  The town of Point-à-Pitre in the former island was entirely overthrown, and 6000 of its people destroyed.  The earth opened in several places, sheets of water spouting from the creeks 100 feet in height; and when the chasms closed, men and other objects were engulfed.  The *eastern side of the United States* is a region where, though earthquakes have been frequent, they are seldom violent.  More than 100 are estimated to have occurred within the past 200 years in the earlier-peopled districts.  The oscillations of the ground in all the Appalachian region have a prevailing north-east and south-west direction.  One of the most conspicuous earthquakes in North America was that which occurred in the *Mississippi Valley* in 1812.  It consisted of a series of concussions, beginning late in 1811, and lasting at intervals for more than a year.  It resulted in the submergence of a vast tract bordering the river on the west.  This tract was permanently altered in level; and some of the chasms opened in the earth during the shocks are still visible.  On 4th January 1843 an earthquake was perceptibly felt from Ohio and Iowa to Louisiana and Georgia, and from the western frontiers of Missouri and Arkansas to the Atlantic coast.  Its duration was estimated at about two minutes, and its rate of motion 33 miles per minute.

## VOLCANOES OR BURNING MOUNTAINS.

These mark certain portions of the earth's surface, where there is a direct communication between the atmosphere outside and highly-heated portions of the interior of the globe.  At such points of communication, *lava*, *ashes*, *scoriæ*, *volcanic dust*, *steam*, and *gases* are emitted.  The gradual accumulation of the solid materials round the orifice gives rise to a volcanic hill.  A volcano is consequently more or less conical in form.  Its summit is usually occupied by a cup-shaped cavity called a *crater* (at the bottom of which lies the *vent* or *focus of eruption*), and its sides are covered with lava, loose blocks of ejected rock, or deposits of cinders and ashes.  Volcanoes are *active*, *dormant*, or *extinct*.  In the active state their craters give out large volumes of steam, with showers of ashes and red-hot stones, while streams of melted lava, escaping either from the crater or from fissures in the sides of the mountain, pour down the slopes, destroying all trees, vineyards, or houses that may chance to lie in their path. A volcano is said to be *dormant* when no eruptions take place from it, but when the escape of steam and gaseous or sulphurous exhalations shows that the subterranean heat can still make itself manifest at the surface. Vesuvius, for instance, had been dormant from time immemorial until in the year 63 it suddenly became active, and destroyed the two Roman cities of Herculaneum and Pompeii. An *extinct* volcano is one where the eruption of lava, ashes, and steam, with the other signs of volcanic activity, has long ceased. In Central France and in Germany there are numerous extinct volcanoes, in which the craters remain quite perfect, though now grass-grown and silent. In the British Islands are found abundantly traces of still more ancient volcanoes, the craters of which have long since been worn away; but the lava and ashes ejected by them remain, and form portions of hills, and sometimes even entire hills.

In Plate III. B are given three illustrations of volcanoes. Fig. 1 is a map or plan of Vesuvius, showing the small modern crater in the centre, with the large partially-effaced ancient crater-wall of Somma half encircling it, and the courses of the lava-streams down the sides of the mountain. Fig. 2 gives a view of Vesuvius from the sea. The modern crater here again occupies the centre, emitting a column of smoke, while the older crater of Somma is seen behind the present active cone. Fig. 3 is a drawing of three extinct volcanoes of Central France. In each of them the hill consists of a cone of ashes, with a central crater, one side of which has been broken down by the current of black lava. In each case the lava has risen within the crater until the pressure burst the weakest part of the crater-wall, and the lava flowed out over the plain.

In some regions large quantities of steam and gas are evolved from small orifices, distinct from ordinary volcano craters. Near Naples, a well-known instance of this kind is called the Solfatara—a name which is now very generally applied to the orifices from which discharges of this kind take place. In some cases the emission of steam and water is effected by sudden efforts, and with great force. Of this nature are the Geysers of Iceland. These are hot springs which for a few minutes throw up a column of water and steam with great violence to a height of 100 feet or more, and then sink into rest for a little till they are ready for a renewed outburst. Fig. 4 shows the eruption of one of these Geysers. Such is the force with which the hot water is driven out that it ascends vertically, even in spite of a strong wind which drives the steam to one side, and allows it to condense and fall in a shower of slanting rain. Stones thrown into the hole from which the water ascends are thrown to great heights, and sometimes even broken to pieces by the violence of the discharge.

## RAISING AND SINKING OF CONTINENTS.

*Movements of Elevation.*—Gradual, and in some cases imperceptible, changes of the relative levels of land and sea are constantly occurring in countries situated at a distance from active volcanoes, and where violent earthquakes are not known. Within the present century the shores of the Baltic, in Scandinavia, have been gradually elevated from 10 to 14 inches above the former level, and are still apparently on the uprise. By the earthquake of Cutch, already referred to, the Ulla Bund, a tract 70 miles in length and 16 in breadth, was upheaved 10 feet, and the Runn of Cutch was sunk so as to create a salt marsh, with an area of nearly 200 miles; and by the earthquake of Chile in 1835, a tract not less than 100,000 square miles was permanently elevated about 6 feet above its former level, and parts of the sea-bottom remained dry at high water, with beds of shellfish adhering to the rocks on which they grew.

There are also evidences of upheaval on the coasts of Denmark, and on the Prussian shores of the Baltic ; and recent observation points to regions of elevation on the north-east shores of Siberia, and at the northern extremity of Baffin Bay. Proofs of the rising of the land are furnished by lines of what are known as " raised beaches." These are terraces of sand and gravel, sometimes containing marine shells, and mark old sea margins. They run along the coast in many parts of the British Islands, at heights of 25, 40, 70 feet, and more, above the present sea-level, each raised beach or terrace indicating a line at which the land on it stood before it reached its present elevation. In Plate III. B, figs. 8 and 9, two illustrations of these features are given. Fig. 8 is a sketch of a remarkable series of ancient beach-lines near the town of St Andrews (*Chambers*). Fig. 9 shows, in a sectional form, how such lines of beach remain on a sloping surface of land. The highest terrace (*a*) is of course the oldest, and marks a time when the sea stood at that level, or the land was so much lower than it is now. The next in descending order is marked *b*, and shows another line at which the sea stood for a long while after the land had risen from *b* to *a*. As a rule, the lower terraces are best preserved, inasmuch as they are younger, and have been for a shorter time exposed to the tear and wear of the elements.

*Movements of Depression.*—A slow submergence of the land is seen on the south-west coast of Greenland, where ancient buildings are now covered by the sea, and where the Moravian settlers have been obliged more than once to move inland the poles to which they attach their boats, the old poles remaining beneath the water. The most remarkable example, however, of the slow sinking of a wide area of the earth's crust is furnished by the coral reefs of the Pacific and Indian Oceans. The general character of a coral islet is shown in fig. 6, Plate III. B. As there indicated, it consists of a low ring of land, with a sheet of calm water inside, and the wild ocean outside. At one time it was supposed that these remarkable circular islands marked the sites of submerged volcanoes. It is now known, however, that they point to a slow sinking of the bed of the ocean. The nature of the evidence which they furnish is illustrated by fig. 7, Plate III. B, which shows three sections, each of which marks a stage in the growth of a coral island. In the first section (A) we see a mass of high land rising out of the sea as a mountainous island. Along its margin reefs of coral are formed by the coral polypes, and are known as *fringing reefs*. A slow subsidence of this part of the earth's crust is, however, in progress. The island sinks gradually, but not too fast to allow the coral reef to be kept up by constant additions to its original height. In this way the distance from the reef to the shore of the island inside increases, and the reef becomes what is known as a *barrier reef* (B). The downward movement still going on, and the coral polypes continuing to keep their building abreast of the waves, the old island at last disappears. On its site there now lies an expanse of calm, pale green water, protected by the circle of reef from the open ocean outside. This central water is the *lagoon*. The sea breaking on the reef detaches fragments of it, and scatters them over the reef, along with the sand and mud produced by the grinding of the fragments. In this way small portions of dry ground are formed, which gradually increase until they more or less encircle the lagoon. Seeds drifted over the ocean come ashore and take root, and thus arise the flat, palm-covered coral islets of the Pacific. The regions in which these coral islands occur point, therefore, to widespread movements of depression of the crust of the earth. They are built, in short, upon the tops of a submerged continent.

The Map (Plate V.) is intended to give a general view of the distribution of earthquakes and volcanoes over the globe, and of the areas which are now, or which have in comparatively recent geological times been, rising or sinking. It will be seen that, on the one hand, earthquakes and movements of upheaval are for the most part found in or not far removed from volcanic regions; while, on the other, the wider areas of submergence are distant from these regions.

## PLATE IV.—Geological Map of the British Isles.

Perhaps no part of the earth's surface comprises within the same extent of surface so complete an epitome of the structure of the rocky crust of our planet as is to be found within the compass of the British Islands. The geological formations which in other countries spread out over hundreds and thousands of square leagues, are here brought within such narrow limits that a single English county contains within itself a greater variety of rocks than can be seen in some whole kingdoms abroad. In this way, without quitting his country, a native of Great Britain may often obtain, from an examination of the rocks around him, a more correct and vivid conception of the nature of geological reasoning, and of the grandeur of geological history, than if he travelled for hundreds of miles in some other countries.

In the Map, to which the following pages are appended as a short explanation, it will be seen that the colours are arranged without much reference to the ordinary geographical subdivisions, and that they cross counties and mountains and rivers, apparently in defiance of all order and symmetry. Each of these colours denotes a distinct geological formation or group of rocks. In order to understand the reason why they are placed in such irregular grouping, it may be well to attend to a few preliminary explanations.

In examining what is known as the *crust of the earth*—that is, the portion of the earth which is accessible to our observation, whether at the surface, or in caves, quarries, pits, and mines—geologists have found that there are two great classes of rocks, one called AQUEOUS or SEDIMENTARY, the other IGNEOUS. Between these two divisions there is an intermediate group, known as the METAMORPHIC, of which the affinities are sometimes closest with the aqueous series, and sometimes with the igneous. Now it is of these rocks that the crust of the earth is formed, and consequently it is these same rocks which we have to study when we wish to become acquainted with the geological history of the British Isles.

The AQUEOUS rocks, as their name denotes, have resulted from the operation of water. In short, they are simply sediment which has been collected under water and deposited there, layer after layer, just as gravel, sand, and mud are accumulated under water in every pool, river, lake, and sea at the present day. The circumstance that this deposition has taken place in layers, seams, beds, or strata, according as the motion of the currents or the supply of sediment varied, has given rise to the term *stratified*, as applied to these successive sheets of different kinds of sediment piled over each other. As the ocean covers by far the larger part of the globe, so the sedimentary accumulations now forming must be chiefly marine, seeing that not only do the tides and currents carry out sediment from the shore, but streams and rivers are everywhere bearing enormous quantities of sand and mud from the land to the sea. So also in past time the greater part of the debris worn away by rains, springs, frosts, rivers, and breakers, from the surface of the land, must have been laid down under the sea. Hence by far the larger number of the subdivisions into which the aqueous rocks have been grouped are of *marine* origin. In further examining the crust of our planet, we find that these aqueous rocks, which were once ordinary gravel, sand, or mud, contain the remains of both plants and animals, just as we see similar remains carried down by rivers or swept out by tides, and deposited among ordinary sedimentary accumulations. But as our investigation advances, we begin to perceive that, in the great majority of cases, neither the plants nor the animals are quite the same as those which live at the present time. And the older the group of aqueous rocks, the further removed are their *fossils* or *organic remains* from the forms of the living world of to-day. Tracing out these differences, we gradually learn that in the rocky crust on which we live and move there is a history of life and change; that from the oldest strata preserved to us up through all the others which have been successively piled over them, there can be traced an onward and upward progression, race after race of plants and animals succeeding each other, approaching nearer and closer to the existing types, until at last we reach the present time, where, as the crowning act of the long history, Man takes the lead in creation.

Arranged in order of time,—that is, in the order in which they have successively been deposited during the past history of the globe,—the aqueous rocks of the British Islands are—

### POST-TERTIARY PERIOD.

| | | |
|---|---|---|
| RECENT. | { | Modern alluvium, peat, and raised beaches. |
| POST-PLIOCENE. | { | Ancient alluvia of the Thames, Ouse, etc.<br>Cave-deposits.<br>Glacial drift. |

### TERTIARY or CAINOZOIC PERIOD.

| | | |
|---|---|---|
| PLIOCENE. | Newer. | Norwich crag. |
| | Older. | Red crag.<br>Coralline crag. |
| MIOCENE. | | Leaf-beds of Mull, basalt of Antrim, lignite of Bovey Tracy. |
| EOCENE. | Upper. | Upper part of fluvio-marine beds of Isle of Wight.<br>Lower part of fluvio-marine beds of Isle of Wight. |
| | Middle. | Bagshot beds.<br>London Clay. |
| | Lower. | Woolwich and Reading beds.<br>Thanet sands. |

### SECONDARY or MESOZOIC PERIOD.

| | | |
|---|---|---|
| CRETACEOUS. | Upper. | Chalk.<br>Upper Greensand and Gault. |
| | Lower. | Lower Greensand. |
| WEALDEN and PURBECK. | | Weald clay and Hastings sand.<br>Purbeck beds. |

**OOLITE or JURASSIC.**

Upper Oolite.
- Portland beds.
- Kimmeridge clay.

Middle Oolite.
- Coral rag, or Coralline oolite.
- Oxford clay.

Lower Oolite.
- Cornbrash and Forest Marble Group.
- Great or Bath oolite.
- Fuller's earth.
- Inferior oolite.

Lias.
- Sands, Upper Lias clay.
- Marlstone.
- Lower Lias clay.

**NEW RED SANDSTONE or TRIASSIC.**

Keuper.
- Penarth or Rhætic beds.
- Red marls, with rock-salt and gypsum.
- White and brown sandstones.

Bunter.
- Upper red and mottled sandstone.
- Pebble beds.
- Lower red and mottled sandstones.

## PRIMARY or PALÆOZOIC PERIOD.

**PERMIAN.**
- Magnesian Limestone.
- Conglomerate, sandstone, and red marl.

**CARBONIFEROUS.**
- Coal-Measures.
- Millstone grit.
- Upper limestone shale and carboniferous limestone.
- Lower limestone shale; calciferous sandstones of Scotland.

**OLD RED SANDSTONE or DEVONIAN.**
- Upper.
- Middle.
- Lower.

**SILURIAN.**

Upper.
- Ludlow beds.
- Wenlock beds.
- Upper Llandovery rocks.

Lower.
- Lower Llandovery rocks.
- Caradoc or Bala beds.
- Llandeilo flags.
- Lingula flags.

**CAMBRIAN.**
- Cambrian sandstones, grits, shales, and conglomerates.

Fundamental or Laurentian Gneiss.

If these various groups and formations had been deposited above each other in regular succession, their united mass would have reached a thickness perhaps twice as great as the height of the loftiest mountain on the surface of the earth. This of course was physically impossible; for if the relative levels of sea and land had remained the same, the land would in time have been entirely worn away, the inequalities in the sea-bed would have been filled up with sediment, and one shoreless ocean would have tumbled round the globe. No more sediment could have been brought to augment the quantity of that which had been already deposited, and hence the total depth of stratified rocks which could have been formed must have been comparatively limited. Instead of an unbroken sequence, however, the stratified portion of the earth's crust presents numerous proofs of interruption. We see how at one time a certain region extended as a wide surface of land; how in after ages it sank beneath the sea, and became covered over with a thick series of marine deposits; how again it rose and received new races of plants and animals, which lived and died on the soil that had gathered over the hardened sand and mud of the old sea-bottom, wherein shells and corals lay entombed by thousands; how perhaps the same district once more settled down beneath the sea, while a new succession of layers of sand and mud, many thousands of feet in thickness, gathered over its submerged forests, until finally it reappeared as dry land, and received its present vegetation and inhabitants. By this means, while the existing proportions in the distribution of sea and land may not have varied much in past time, there has been, nevertheless, a constant inter-change of land and sea over the surface of the globe, one part sinking beneath the ocean, another rising above it. Although, therefore, one unbroken deposition of sedimentary rocks has not been going on continually over the whole surface of the globe, there has yet been always a deposition in progress somewhere. It is the task of the geologist to piece together such fragmentary remains of these scattered strata as may have been preserved, to gather up the evidence which they may furnish as to the progression of life and of terrestrial change, and thus to arrive at some conception of the past history of our planet.

It has just been said that the stratified rocks afford abundant proofs of interrupted succes-sion—that is, they show that a group of strata which was in course of formation at the sea-bottom was upheaved, exposed to denudation, worn down to a greater or less extent, and at last covered over with another later series of deposits. This upheaving of rocks is an important element in the geological economy of nature. It points to the existence of an expansive force lodged within the crust of the earth, manifesting itself sometimes in the form of slow movements of uprise or depression, sometimes in sharp earthquake shocks, and sometimes in the action of volcanoes. These subterranean movements are the origin of the second or IGNEOUS group of rocks, which include lava, pumice, scoriæ, and the other products of an active volcano, together with a large number of rocks found in the crust of the earth, and whose existence there is to be traced to the ejection of mineral matter from some highly-heated source below.

METAMORPHIC rocks are strata which have been subjected to a process of change within the crust of the earth, whereby they have been altered from the condition of sandstones, shales, and other ordinary sedimentary deposits, into hard crystalline masses, in which the signs of stratification are more or less effaced. The cause of this change or *metamorphism* is not yet well understood. It can be traced in all stages of its progress from common sandstone and shale into greywacke and slate, thence into various *schists*, as *mica-schist, talc-schist, chlorite-schist*, according to the mineral composition of the rock, thence into *gneiss*, until at last we reach granite, the most crystalline and amorphous of the whole, and apparently the ultimate stage of the process. Whatever may have been the origin and mode of this change, it probably went on far below the surface; and in some cases, though possibly not in all, it was accompanied by such a degree of heat as fused the rocks below, and injected the molten mass into every crack and fissure into which it could find its way. Hence in some of their relations the metamorphic are closely connected with the igneous rocks. Heated water, charged with alkaline or other solutions, has perhaps had a large share in the process of metamorphism.

Each of these three classes of rocks is abundantly developed in the British Islands.* The stratified class covers, of course, by far the largest extent of surface. The igneous group is chiefly developed along the great central valley of Scotland, and along the west coast northwards from the basaltic plateau of Antrim. But igneous rocks occur also abundantly in other districts, as shown on the Map. They appear to be wholly absent, however, from the southeastern half of England. A line drawn from the Yorkshire coast about Scarborough to the mouth of the Severn divides the kingdom into two nearly equal parts, whereof that to the northwest contains numerous local protrusions of igneous rocks, while that to the south-east shows none. The metamorphic series finds its great development in the Highlands of Scotland, but it also occupies a considerable space in the north of Ireland. Metamorphism is likewise seen round the granite bosses in the Lower Silurian hills of the south of Scotland, among the Lower Silurian rocks of Anglesea, among the Devonian rocks of Cornwall and Devon, and in the Carboniferous series round the granite of Dartmoor.

By comparing this map with an ordinary geographical one of the same region, it will be seen that certain formations occur in the mountainous and hilly districts, while others are confined to the lowlands and plains. A little attention to this subject will soon show that there is a close connection between the external form of the ground and the nature of the rocks underneath the surface. As a general rule, the high parts of the country consist of hard rocks belonging to the oldest or *palæozoic* division, while the low grounds lie usually upon soft strata of the later or *secondary* and *tertiary* groups. The Highlands of Scotland, for example, are made up of hard tough crystalline rocks, such as gneiss, mica-schist, quartz-rock, granite, and porphyry. The mountains of Wales and Cumberland consist of compact Silurian sandstones and grits, with thick beds of old volcanic lavas and tuffs. In Ireland, too, the Wicklow chain of mountains is one of granitic rocks. The great plains of central and south-eastern England, on the other hand, lie upon the comparatively soft sandstones, marls, and clays of the new red sandstone, oolitic, and eocene formations.

### PLATES VI. to X.—The Mountains and Table-Lands of Europe, Asia, Africa, and America.

This series of maps is intended to convey correct information regarding the *relief* of the continents. The *physical* position of a place, or its elevation in the atmosphere above the level of the sea, is an element of as much importance in regard to its climate, health, vegetation, and other circumstances, as its *geographical* position, or its vicinity to, or distance from, the Equator. This information on the physical or natural geography of a country cannot be conveyed in an ordinary geographical map, which must necessarily contain many names and political divisions, to the exclusion or obliteration of purely physical features. The forms of relief on the surface of the globe are varied to an almost infinite extent, yet they may be classified according to their most prominent characteristics.

* It has not been deemed desirable to describe here the distribution of the various formations across the country, as this is sufficiently indicated upon the map. The nature of the formations themselves, with their characteristic fossils, may be learned from a geological text-book.

These are—1. *Mountains*, which, disposed in linear chains or scattered in isolated groups, extend over the surface of plains and table-lands in every variety of form.

2. *Table-lands* or *plateaus*—elevations of the surface rising to a considerable height, and presenting the form of a table or platform; and—

3. *Plains* and *valleys*, or *lowlands*—portions of the surface little or not at all raised above the level of the ocean.

On comparing the great mountain-chains of the different continents, the first observation which occurs is, that in the Old World—Europe, Asia, and Africa—they follow, in general, the direction of the parallels of latitude, or extend nearly in a direction from east to west; while in the New World, North and South America, the principal chains extend in the direction of the meridians, or from north to south. And next, that in all cases the table-lands are intimately connected with the mountain-ranges—the highest mountains invariably rising, not from plains, but from table-lands. The Old World is most remarkable for its mountains and table-lands, and the New World for its plains.

EUROPE, a peninsular prolongation of Asia, of which it forms the western portion, presents a surface differing in elevation from the plains of the Netherlands, at or below the level of the sea, to the Alps, nearly 16,000 feet above that level. The great mass of the Continent is divided into a region of mountains and high lands in the south-west, and one of low lands and plains in the north and east. The mountainous girdle extends from the Atlantic eastward to the Black Sea, attaining its greatest elevation in *Mont Blanc*, 15,810 feet above the sea-level, the highest point of the *Alps* and of Western Europe.* The Alps constitute the centre of High Europe, and form a network of wild and broken ranges, in which are enclosed narrow but fertile valleys. The tops of the mountains are covered with perennial snow, and from them glaciers descend to the plains. The *Apennines* stretch from the southern part of the Alps south-east throughout the entire length of Italy, forming a long narrow chain from which wide valleys depend. The highest point of the range is the Gran Sasso d'Italia, 10,306 feet above the sea. The *Julian* and *Dinaric Alps* unite the Apennines with the *Balkan* in Turkey and the *Pindus* range in Greece. The *Carpathians* extend from the Eastern Alps in a semicircular form, enclosing the plain of Hungary on the north and east, terminating in the Transylvanian Mountains in the south. The Tatra group in the north rises to 8779 feet above the sea. South-west of the Alps the *Pyrenees*, in which Pic Nethou is 11,168 feet above the sea, form a natural barrier between France and Spain, and are continued westward to the Bay of Biscay by the lower range of the Cantabrian Mountains. The other mountains of the Spanish peninsula are the sierras of Guadarrama and Estrella, near the centre; Morena and Nevada, the most elevated range in the peninsula, Mulahacen being 11,678 feet above the sea, in the south.

The secondary mountains of Central Europe are the *Cevennes*, in the south-east of France; the Mountains of *Auvergne*, with the volcanic group of *Cantal* near the centre; the *Jura* Mountains, between France and Switzerland on the north-east; the *Vosges* on the west and the *Schwarzwald* on the east of the Rhine; the Little *Carpathians*, between the rivers Waag and March, in the east; and the Harz, of which the Brocken, 3740 feet, is the highest point; the *Erz Gebirge* or Ore Mountains, between the Elbe and Eger, and the *Riesen Gebirge* or Giant Mountains, in which Schneekoppe (the Snowy Peak) rises to 5273 feet, in the north of Germany. Separated from these central groups by the low ranges of the *Hundsruck* and the *Taunus* on the north-east and south-west of the Rhine, and the great plain of France, the

* On the south-eastern verge of Europe, within the political boundary of the Continent, two mountain-peaks, Elburs and Kasbek, on the northern slope of the Caucasus range, are respectively much higher than the culminating point of the Alps, but Mont Blanc is the monarch among the mountains of historical Europe.

Netherlands, Denmark, and the North Sea, are the mountains of Scotland, the culminating point of which, and of all Britain, is Ben Nevis, in the Grampians, 4406 feet high; and Wales, in which Snowden, the highest point in South Britain, is 3590 feet above the sea; the Alps of Scandinavia, forming a huge barrier of plateaus and mountain-ranges fronting the Atlantic on the north-west, and the Ural Mountains on the east. The portions of mountains covered by perennial snow in Europe are limited to the more elevated regions of the Alps, the Scandinavian Mountains, the Pyrenees, and the Sierra Nevada. They are indicated by being left white on the brown ground These mark the highest points of a system, the sites of true glaciers, and the sources of important rivers.

The TABLE-LANDS are, in Spain, the *Plateau* of *Old Castile and Leon* in the north, and that of *New Castile* and *Estremadura* in the south: the northernmost of these is the largest and most elevated in Europe, its average height being 2700, and its highest point 4500 feet above the sea. The *Plateau of Bavaria* forms a terrace on the north side of the Alps, divided into two parts by the Lake of Constance. The only others are the *Plateau of Auvergne*, with *Cantal*, in France; and that of *Bohemia*, an elevated basin, 960 feet above the sea, nearly surrounded by mountains.

The great PLAINS of northern and eastern Europe extend from the Atlantic to the Ural Mountains, and from the Caucasus to the Arctic Ocean. So uniform is the surface here that the entire space may be traversed without changing the level more than a few hundred feet. The Valdai Hills, near the sources of the Volga, 1100 feet, are the highest parts in the interior.

The *Lowlands of France* on the west merge into the *Germanic Plain*, including the Netherlands and the plain of Denmark; and this is extended eastwards through Russia by the Sarmatian Plain in the centre, and the plains of the White Sea on the north, and those of the Black Sea and the Caspian on the south.

The SECONDARY PLAINS are, in Spain, the basin of the *Ebro* in the north-east, and that of the *Guadalquivir* in the south-west; in France, the valley of the *Lower Rhine*; in Italy, the *Plain of Lombardy* and the *Pontine Marshes*; in Germany, the *Valley* of the Lower Danube, the *Hungarian* and the *Wallachian* plains; and in Scandinavia, the eastern shores of Sweden sloping to the Baltic and the Gulf of Bothnia. In Great Britain are the *Central Plain*, the valley of the *Severn*, and *Plain of York*, in England; and *Strathmore* and the *Carse of Falkirk* in Scotland. In Ireland, the *Great Plain* occupies the greater part of the interior of the island.

The CLIMATE of Europe is indicated on the map by the annual isotherm of 41° Fahr., and the monthly isotherm for January, and by lines tracing the limits of some of the more characteristic trees, grains, and fruits.

ASIA is the largest of the continents or great divisions of the globe, comprehending all the countries east of Europe and Northern Africa. A central mass of continent forms nearly four-fifths of the entire area, and from this, extensive peninsulas project on the east, south, and west. Like Europe, it is divided into a vast elevated mountainous region, and a region of valleys or plains. 1. The table-land of Central Asia forms an upheaved region in the middle of the continent, and is crowned by the great wilderness called Gobi or Shamo; its southern border extends from the Indus to the eastern shores of China, supporting the Himalayas, which are prolonged eastwards by the Chinese mountains south of the Yang-tse-Kiang. The eastern border of the table-land extends from the Yang-tse-Kiang on the south to the Amoor on the north; the northern border is girdled by ranges of mountains, of which the Altai is the nucleus; and the western edge forms the high land of Turkestan, on which rises the *Bolor Tagh*, called by the natives the *Roof of the World*, the crests of which are supposed to be 19,000 feet above the sea. This table-land is crossed in an easterly and westerly direction—first, by the *Thian Shan* or Celestial Mountains, on which is the high peak of *Peshan;* second, by the *Kuen-lun*, which is extended eastward by the *Peling* Mountains ; and between these is the plateau of Upper Tartary. Between the Kuen-lun and the Himalaya proper is the *Table-land of Tibet*, which appears to be composed of range after range of snow-clad mountains, connected with, and not much inferior in elevation to, the Himalayas.

The mountain system of the *Altai* and its offsets separate Mongolia on the south from Siberia on the north. Extending north-east from the Altai, the mountains of Dauria are prolonged by the Yablonoi Mountains to the Sea of Okhotsk, and the Stanovoi and Aldan Mountains stretch north-west and north-east to the arctic circle. The table-land of *Iran* or *Persia* is connected with that of Central Asia by the Hindoo Koosh Mountains : it has an average height of 2500 to 3500 feet, and on its north side the Volcano of Demavend in the Elburz range is 18,464 feet in elevation. This is joined on the west by the plateau of *Armenia*, about 7000 feet high, supporting Mount Ararat, on which the ark is supposed to have rested after the Flood, 16,964 feet above the sea ; and that of *Asia Minor*, a mountainous mass with a mean elevation of about 3280 feet, on the south side of which are the Taurus Mountains, 9800 feet, and Mount Arjish, 13,197 feet above the sea. The *Table-land of Arabia* appears to have a general elevation of 2000 to 4000 feet ; on the south-east, the mountains are 5000 feet above the sea-level : a large portion of the interior is an arid desert. The *Mountains of Syria and Palestine* form a connecting link between the table-land of Arabia and the Taurus ; the highest point of Lebanon has an elevation of 10,061 feet above the sea.

The mountain system of *Hindostan:* on the north the peninsula is bounded by the Himalayas ; and on the south, the Western and Eastern Ghauts, running parallel with the shores, enclose the table-land of the Deccan, which is bounded northwards by the Vindhya Mountains. The *Himalaya* is not a continuous mountain-chain, but a group of snow-clad peaks widely separated from each other—the spurs of a much broader region of snow-clad peaks, glaciers, and valleys on the north : their mean elevation is probably 16,000 to 18,000 feet. The highest peaks are Dhawalagiri, 26,826 feet ; Kanchinjanja, 28,156 feet ; and Mount Everest, situated nearly midway between them, 29,002 feet, or 5½ miles, above the sea, the highest known point on the surface of the globe. The table-land of the Deccan has a mean elevation of 1500 to 2500 feet above the sea, and is covered with isolated conical hills rising 2000 feet above the plateau. The *Mahabaleshwar Hills* in the Western Ghauts are 4700 feet above the level of the sea. In Ceylon, Pedrotallagalla is 8280 and Adam's Peak 7420 feet above the sea.

The *Plains* of Asia comprise a large portion of true steppes and deserts, interspersed with fertile valleys. The *Chinese Lowland,* a district of pure alluvium, extends from the shores of the Yellow Sea to a great distance inland, and occupies the deltas of the Yang-tse-Kiang and the Hoang-ho. The plains of *Further India* include the rich valleys on the west of the Gulf of Tonquin, and those of the rivers Mekong, Menam, Saluen, and Irrawady. The *Plains of Hindostan* comprise, on the east, the valley of the Ganges, which, at its mouth in the Bay of Bengal, forms a cluster of marshy islands called the Sundarbans. On the west, the *Valley of the Indus*, an alluvial plain, extends for many hundred miles without any eminence, except the salt range of hills near the western base of the Himalaya. In the south it is mostly a desert, called the Thar, and it contains the remarkable depression of the Runn of Cutch (Ran of Kacch). The lowland of *Turan and Bucharia* extends from the table-land of Iran on the south to the Ural Mountains and a western branch of the Altai on the north ; its lower portion contains the great depression of the Caspian and Aral seas, forming the *basin of the Continental streams*, the largest hollow on the surface of the globe, and apparently the bed of a former inland sea. The *Plains and Steppes of Siberia* extend from the Ural Mountains on the west to Behring Strait on the east, sloping northward from the base of the Altai and the Stanovoi mountains on the south to the Arctic Ocean on the north. In the southern regions the surface is covered with forests, and has fertile valleys; in the west the steppes are flat and undulating; and in the north it is a trackless wilderness, with numerous salt lakes and marshy flats, called *Tundras*, the soil of which is frozen to a depth of several hundred feet during great part of the year.

The *Lowlands of Syria and Arabia* are in the south dry and stony deserts, but the lower courses of the Tigris and Euphrates in the north are rich and fertile. In the south of Palestine, the Dead Sea is the deepest and most remarkable depression on the face of the globe, its surface being 1298 feet below the Mediterranean.

**AFRICA** forms an immense peninsula, joined to Asia by the Isthmus of Suez, and separated from Europe by the narrow Strait of Gibraltar and the Mediterranean Sea. Unlike the continents of Asia and Europe, in which the mountains and highlands are grouped near their respective centres, Africa has its principal elevations on its outer margin, and its plains and valleys in the interior. On the north-west coast the Atlas and its ramifications cover a large portion of Marocco and Algeria to Tunis, rising in Mount Miltsin to 11,400 feet above the sea. From Tunis to the delta of the Nile the shores are low, except at Jebel Acdar, east of the 20th meridian. From the Nile delta a series of terrace-formed rocky hills extends southward along the shores of the Red Sea to the alpine region of Abyssinia, a confused mass of elevated table-lands topped by high mountains, among which Ras Detschen is 15,986 and Abba Jarrat 15,020 feet above the sea-level. South-east from Abyssinia to the equator the country continues high on the eastern border, and the north shore of the Somauli country is 6500 feet above the sea. South of the equator the volcanic peaks of Kenia and Kilimandjaro mountains are supposed to be 20,000 feet above the level of the sea, and are snow-clad.

The mountainous or hilly margin continues southward, but of uncertain elevation, and to an unknown extent inland. Where broken through by the Zambeze, the Lupata Mountains are only from 600 to 800 feet high, but farther north, between Lakes Nyassi and Shirwa, they rise to 7000 feet. South of the Zambeze the country rises in the Quotlaanba and Drakenberg mountains to 10,317 feet high in Natal. The Compass Berg, the highest point in the Cape Colony, is 8500 feet above the sea. The highland is now continued west and north-west along the north of the Cape Colony by the Sneeuw Bergen (Snowy Mountains), the Nieuweveld and the Roggeveld mountains, north of the Karoo plain (an arid tract, 200 miles long and 50 miles broad), towards the Orange River. North of this the mountain-zone occupies a space between the coast and the Kalahari Desert; and in Damara Land, Omatako Mountain rises to 8739 feet. The interior of the southern portion of Africa has a general elevation of from 3000 to 4000 feet, Lake N'gami in the centre being 3285 feet above the sea-level. An elevated table-land extends the mountainous fringe north to the Bight of Biafra, where, in the volcanic group of the Camaroons Mountains, Albert Peak, an enormous crater, is 13,000 feet above the sea. Beyond the delta of the Niger, King William Mountains, 2000 to 3000 feet high, are continued westwards by the Soracte Mountains, 1278 feet; Rennel Mountains, 3200 feet; and the Kong Mountains, 2000 to 3000 feet; and then the high border-land terminates abruptly in Senegambia, on the south edge of the Sahara.

In the interior of northern and eastern Africa, all preconceived opinions have been changed by recent discoveries. The Sahara is not, as was believed, a low sandy plain, but—at least in its central and eastern portion, explored by Barth and Duveyrier—a desert traversed by hills and table-lands, and interspersed with fertile oases. South of Tripoli the surface varies from 1000 feet till the desert rises to its highest point, south-west of Mourzouk, in a plateau 4000 to 5000 feet above the sea. In the high land of Soudan, south of Lake Tchad, Mount Mindif is 6000 feet and Mount Alantika 9000 feet above the sea.

The interior of South Central Africa, as predicted by Murchison and ascertained by Livingstone, has the form of an elevated trough, the outer fringe of rocks already described enclosing an immense region, well watered by streams and lakes. The Zambeze, which is understood to rise north of the parallel of 10° S., near the centre of the peninsula, flows south through the low Barotse valley to the 18th parallel, when, breaking through a watershed 5000 feet high at the Victoria Falls, it flows east and south-east to the Indian Ocean. The basin of the Zambeze, the form of which is determined by the high lands on the east, west, and south, is traversed by numerous tributary streams. The eastern region south of the equator, explored by Burton and Speke, rises from the coast plains to fertile ridges and mountain groups, succeeded by an elevated plateau, west of which the land sinks to the lake region, where, on the equator, Victoria Nyanza Lake, 3308 feet above the sea, is a principal reservoir of the White Nile, north-west of which the fine lake Albert Nyanza is nearly on the same level. South-west of this, between lat. 3° and 8° S., Lake Tanganyika, 300 miles long and 1800 feet above the sea, is fed by streams from mountains 6000 to 8000 feet in elevation.

There are no extensive low-lying plains in Africa, the low ground being limited to certain parts of the coast, near the river deltas, as in that of the Nile, the coast-streams of Senegambia, the mouths of the Quorra, and those of the Zambeze. In the interior, the basin of Lake Tchad, 830 feet above the sea, is an extensive tract of valley-land; and in the east, between Lake Tanganyika and the Indian Ocean, is a vast level plateau, 2000 to 4000 feet above the sea.

**NORTH AMERICA** comprises all that portion of the American continent north of the Isthmus of Panama; its area is more than twice that of Europe, and more than one-sixth of all the dry land of the globe. Its axis of elevation is on its western side, whence its surface slopes to the east, north, and south-east. A vast series of table-lands, more than 3000 feet in mean elevation above the sea, forms a broad irregular belt on the Pacific side, extending from Behring Strait to the Caribbean Sea. Along the western edge of this plateau, from the territory of Alaska to the peninsula of Lower California, stretches the *Oceanic Mountain Chain*, parallel to the Pacific. At its northern extremity are the volcanic peaks of *St Elias* and *Fairweather*, rising to nearly 15,000 feet. South of this the *Sea Alps* are succeeded by the *Cascade Range*, where *Mount Baker*, an active volcano, is 11,000 or 12,000 feet, *Mount St Helens* 14,000, *Mount Hood* 12,000, and *Mount Shasta Butte* 14,440 feet above the sea. Between the parallels of 35° and 40° the *Coast Range* and the *Sierra Nevada* separate to enclose the fertile gold-producing valley of California in the basins of the Sacramento and Joaquin rivers. Several peaks of the Sierra Nevada attain a height of 10,000 feet. *Mount Dana* is 13,000 feet above the sea. South of this the mountains decline in elevation, through the Sierra de Santa Lucia (in which, however, Mount Whitney is 15,000 feet high) to Lower California.

The *Rocky Mountain Chain* forms the principal watershed of the continent, all the waters west of this flowing to the Pacific, and all east to the Arctic and Atlantic Oceans. It comprises two, and in some places three, distinct ranges of high mountain-crests, table-lands, and valleys. The main chain forms a waving line curving along the central and eastern part of the table-land; it is not continuous, but broken by deep passes, and the leading ridges often overlap each other. Its eastern range contains mountain-peaks rising 10,000 and 12,000 feet above the sea. North of the River Platte, the *Wind River Mountains*, the highest of the chain, rise in *Fremont's Peak* to 13,576 feet. This lofty axis separates the head-waters of the Missouri, flowing to the Atlantic, from those of the Columbia, which flow westward to the Pacific Ocean. North of the Wind River chain, the eastern range, near the sources of the Saskatchewan, rises in Mount Hooker to 16,750 feet, Mount Murchison to 15,789 feet, and Mount Brown to 16,000 feet above the sea. From this point the mountains of the eastern range decline northwards to the Arctic Ocean. South of the Wind River Mountains the eastern ridge is marked by *Union Peak; Long's Peak*, 15,000 feet; and *James's Peak* and *Pike's Peak*, 14,216 feet; and continued southward to the *Cordillera of Coahuela* and *Potosi*, forming the eastern edge of the Mexican table-land. Its western ridges are the *Sierras Verde, Mimbres,* and *Madre,* enclosing on this side the *Table-land of Mexico or Anahuac,* which has an elevation of from 6000 to 9000 feet above the sea. The plateau is in some places traversed by well-defined ridges, but in others it is unbroken either by hills or depressions. The city of Mexico is 7473 feet above the sea, and on the south-eastern and loftiest side of the table-land the volcano of *Popocatepetl,* the highest point of the Mexican system, is 17,720 feet high. East of this is the volcano of *Orizaba,* 17,374 feet above the sea. The *Volcanic Range of Guatemala* comprises a great number of volcanic peaks, rising from a table-land about 5000 feet above the sea: of these, *Agua,* 13,000 feet in elevation, is verdant to the summit. Immediately west of it the volcano of Fuego is 13,800 feet high.

Between the Rocky Mountains on the east and the coast-chain and Sierra Nevada on the west there extends, from the Gulf of California to the Arctic Ocean, a high desert zone, which in its central regions, between lat. 35° and 45° N. in the *Great Basin of Utah,* is from 4000 to 5000 feet above the sea. This region is nearly rainless, and comprises wide plains encrusted with salt. It contains twelve or fifteen saline lakes, the largest of which is the Great Salt Lake of Utah, 4238 feet above the sea.

The *Mountains of Veragua* are upwards of 5000 feet above the sea-level, and the *Isthmus of Panama* is about 850 feet in mean elevation, but the Pacific Railway crosses at a height of only 262 feet above the sea.

The *Appalachian or Atlantic Mountain System* (Alleghanies), on the east, is the only counterpoise to the vast mountain-mass on the west side of the continent. It embraces the entire range of high lands extending from the Gulf of St Lawrence to Georgia, about 1000 miles in length, and 150 to 200 miles in breadth. It is composed of many parallel ridges, with crests reaching to 6700 feet, their bases resting on a plateau, the mean elevation of which is 2500 feet above the sea. At the south end of the Blue Ridge, Mount Black is 6707 feet high.

The *Great Central Plain* of North America extends northward from the Gulf of Mexico to the Arctic Ocean, between the Rocky Mountains and the Appalachian chain. It is a vast region of plains, slopes, and table-lands, watered by many great rivers, and indented by large lakes; on the west it forms a region of dry elevated steppes or table-lands, and on the east a wider and lower tract of hills and rolling plains, generally well watered. Near the middle of the plain a broad gentle swell of the land crosses the continent from the coast of Labrador, north of Lake Superior, west-south-west towards the Pacific coast. In the *Missabay Heights,* west of Lake Superior, it is 1500 feet above the sea, and divides the waters flowing north to the Arctic Ocean from those which flow south to the Gulf of Mexico.

The *Atlantic Slope* stretches from the Gulf of St Lawrence to the Gulf of Mexico, narrow in the north, but expanding southwards to 200 miles in width from Virginia to the southern end of the Alleghanies; its greatest height is about 1000 feet, near the sources of the Roanoke. The *Atlantic Plain,* or seaboard belt of the Atlantic slope, is nowhere more than

100 feet above the level of the sea; in the south-west its seaward portion is swampy and overflowed. This lower plain continues along the shores of the Gulf of Mexico, the peninsula of Yucatan, and the shores of Honduras, terminating in the valley of the San Juan and the Lake of Nicaragua.

The continent of North America comprises a broad *forest zone* in the south-east, a narrow zone of forests on the Pacific coast, and a vast *treeless region* extending from the Pacific-coast mountains to the Mississippi. The northern limit of trees cuts the point of Labrador, and passes north of Great Slave Lake. North of this is the *Arctic treeless region*. On the map are marked the northern limits of the growth of wheat, and of the wine-yielding vine, as well as the annual isothermals of 40°, 50°, and 60° Fahrenheit.

In **SOUTH AMERICA**, as in the northern portion of the continent, the elevated axis is on the west or Pacific side, where the great mountain-system of the Andes stretches, through 65° of latitude, from Cape Horn to the Isthmus of Panama, with a breadth of 40 to 400 miles. The centre of the continent is occupied by the immense plains of the Amazon, the Plata, and the Orinoco; and in the east are the separate mountain-systems of Parimé and those of Brazil. The *Cordillera* or *Great Chain of the Andes* is divided, according to the countries which it traverses, into five principal sections or groups: (1.) The *Andes of Quito*, extending from lat. 5° S. to the Isthmus of Panama in lat. 8° N. In this division more of the giant mountains are congregated than in any other part of the chain. On the east side the snow-clad summit of *Antisana* is 19,137 feet, and the beautiful white cone of *Cayambe*, traversed by the equator, 19,535 feet above the sea; and on the west, *Pichincha* is 15,924 feet in elevation. Between these ranges lies the table-land of *Quito*, with the city of Quito, 9543 feet above the Pacific. In the chain of *Quindiu* is the peak of *Tolima*, 18,020 feet, the highest point of the Andes north of the equator. (2.) The *Andes of Peru* extend from lat. 15° to 5° S. Near the south-east extremity is the mountain-knot of *Vilcakota*, 17,525 feet. In about lat. 10° is the mountain-knot of *Huanuco* and *Pasco*, 11,800 feet, on which rises the Nevada de la *Viuda*, 16,000 feet. (3.) The *Andes of Bolivia*, the central and most elevated portion of the system between lat. 21° and 15° S., consist of two great parallel mountain-chains—the Cordillera of the coast, and the eastern or Bolivian Cordillera—enclosing a vast Alpine valley, the plateau of Bolivia, partly occupied by the lake of Titicaca, 12,847 feet above the sea. In the western chain are the Nevada of *Gualateri*, 21,960 feet; *Sahama*, 22,350 feet; the volcano of *Arequipa*, a regular cone, 20,320 feet; and the Nevada of *Chuquibamba*, 21,000 feet above the sea. The eastern chain, or the *Cordillera Real*, has a mean elevation of 16,000 feet. In it rises the nevada of *Illimani*, with an elevation of 21,140 feet, the lowest glaciers on the northern side of which do not descend below 16,500 feet. North-west of this is the Nevada de *Sorata*, or *Ancohuma*, the most elevated of all this snow-capped range, 24,812 feet. (4.) The *Andes of Chile* extend from lat. 21° to 42° S., running for a considerable extent in a single ridge of about 30 miles in breadth. The mean elevation of the chain is about 12,000 feet, but several serrated peaks rise to a much greater elevation. Visible from the port of Valparaiso rises the giant of the Chilean Andes, the *peak of Aconcagua*, 23,301 feet above the sea-level. Inland from this, near the capital of Chile, is the nevada of *Tupungato*, 15,000 feet. There appear to be five active volcanic vents in the southern prolongation of the Andes of Chile—viz., *Maypu*, *Chillan*, *Antuco*, *Villa Rica*, and *Osorno*. (5.) The *Patagonian Andes* commence at the Strait of Magellan, and run north to lat. 42°, with an elevation of 3000 to 8000 feet above the sea. Their summits are covered with perennial snow, whence glaciers descend almost to the sea-shore. The chief of these snowy mountains are *Osorno*, 7550 feet above the sea; *Yanteles*, 8030; *Minchinmadom*, 7400; and *Stokes*, 6400 feet.

Under the parallel of lat. 21° S. the snow-line rises to 17,000 feet; in lat. 26° 30' it is 13,800 feet, and on the parallel of Valdivia it sinks to 8300 feet.

The *Mountain System of Parimé* forms a plateau between the Orinoco, the Rio Negro, and the Amazon; the principal chains run in a direction generally from east to west. Near the centre the sierra *Pacaraima* is 7000 feet, and on the west, *Mount Maravaca* is 10,500 feet above the sea-level.

The *Mountains of Brazil* form a vast isolated plateau between the Amazon, the Madeira, the Paraguay, and the Plata rivers. In the south the *Serra do Mar*, or coast-range, forms the south-east edge of the plateau. This, with the *Serra dos Orgaos*, or Organ Mountains (so named from the resemblance of their peaks to the tubes of an organ), compose the so-called Brazilian Andes. Proceeding northward, the other chief ranges are the *Serra do Espinhaço, Serra Timba, Serra Tabatinga*, and *Serra Irmaos.* The *Cordillera Grande* and the *Serra Santa Martha* divide the waters of the Araguay and the Tocantins, and a high ridge south of the Orinoco separates Brazil from Venezuela. The highest points of the Organ Mountains reach an elevation of 7500 feet, and Mount *Itambe* is 5755 feet above the sea-level. South of this the *Serra de Piedade* is 5830 feet, and *Itacolumi* 5750 feet in elevation. Immediately within the coast-range, the *Sertas*, or table-land of Brazil, rises by several gradations to the central plateau, the mean height of which is estimated at from 2500 to 3000 feet.

The PLAINS of SOUTH AMERICA comprise three principal and several subordinate divisions. (1.) The *Llanos* or *Plains* of the *Orinoco*, between the mountains of Brazil on the east and the chain of Suma Paz on the west. These plains are perfectly level, and at a distance of 450 miles from the sea-coast the surface is only 192 feet above the sea. Being situated within the torrid zone, they are, during one-half the year, desolate sandy wastes, while during the other half they are covered with luxuriant grass. (2.) The *Atlantic Slope*, the *Valleys of the Amazon* and of the *Rio de la Plata.* All the large rivers of South America flow eastward to the Atlantic Ocean, indicating the slope of the surface in this direction. The valley of the Amazon, the largest river-basin in the world, is situated partly in the northern, but mostly in the southern hemisphere. Its climate forms a striking contrast to that of the Orinoco, for it is copiously watered at all seasons, supporting a luxuriant vegetation, and its humid plains are occupied by impenetrable forests, the growth of a thousand years.

The *Valley of the Rio de la Plata* consists chiefly of two immense plains, the northern watered by the Salado, Vermejo, Pilcomayo, &c., and the southern comprising nearly all the *pampas of Buenos Ayres*—a vast plain, covered for the greater part with rich pasturage and gigantic thistles, interspersed with saline lakes.

The *Patagonian Steppe* is composed of a series of terraces east of the Andes, covered with marine shells, clay, earth, and gravel: these terrace-slopes are arid and sterile.

The *Pacific Slope* is a narrow belt between the chain of the Andes and the ocean, from 50 to 100 miles in breadth. In the south it comprises the moist shores and islands of Patagonia, in Chile the fertile provinces of Talca, San Fernando, and Conception, and from lat. 30° to lat. 10° S. the rainless district of Peru. Its fertile plains are frequently interspersed with vast deserts of sand.

### PLATE XI.—Ocean Currents,

gives a connected view of the ocean currents over the globe. These singular marine movements are distinguished as *constant, periodical,* and *variable* currents—the latter two being produced by winds and tides. The constant or true ocean currents commence at the South Pole, under the name of the *Antarctic Drift current*, which, after pouring a stream of cold water along the shores of Chile and Peru, flows westwards through the Pacific and Indian Oceans as the *Equatorial current*, entering the Atlantic by the *Cape current*. The direction of the stream is now northwards along the western shores of Africa, till, near the Equator, it is carried westward by the *Equatorial current* of the Atlantic, which, entering the Gulf of Mexico, originates the *Gulf stream*. This remarkable current carries an immense volume of warm water from the Gulf of Mexico across the North Atlantic, in a north-easterly direction, towards the shores of Britain, France, Norway, and even to Iceland and Spitzbergen. The Gulf stream derives its name from the Gulf of Mexico, but has its origin in the Great Equatorial current. After imbibing the heat which accumulates in the gulf, it escapes eastward through the Strait of Florida. When off the southern shores of the United States, its surface temperature is 84° Fahr., and its rate of motion 80 miles in twenty-four hours. It soon begins to expand, when its velocity decreases to 60, 40, and 30 miles, as shown on the Map, and the temperature of the water is gradually reduced. Ocean currents are of the greatest service in the distribution of temperature; some, as the Peruvian and the Carribean currents of cold water, carrying off the superfluous heat from countries which would otherwise be nearly uninhabitable; while others, as the Gulf stream and the Japan current, convey heat and moisture to the cold and desolate regions of the north. Some of them maintain a constant difference of many degrees between their own temperature and that of the water in their immediate vicinity. The north-east branch of the Gulf stream carries warm water to the shores of the British Islands, and raises our winter temperature far above that of Labrador in corresponding latitudes. Its influence on the coast of Norway is so great that even up to North Cape, lat. 71°, its tepid waters keep the harbours open all the year round; while the Baltic, many degrees farther south, is frozen during the winter months. The great counter-current of the Gulf stream is the *Arctic current* of cold water from the polar regions, which, sweeping down the eastern shores of Greenland, passes Cape Farewell, and unites with the Labrador current from Davis Strait. The united stream, flowing southward with its freight of icebergs burdened with the debris of splintered rocks, meets the northward-bound water of the Gulf stream in the region of Newfoundland, where the result of the conflict that ensues between the cold and heated waters is the melting of the greater portion of the ice, and the deposit of a vast amount of soil, which, in the course of ages, has formed the famous codfishing banks of Newfoundland; and where the opposing currents of the atmosphere produce the beautiful but treacherous silver fogs, so much dreaded by the mariner. Currents are also very important to navigation: a striking example of this is exhibited in the *Equatorial* and *Guinea currents* of the Atlantic. The first flows north-west, and the second south-east; they are in close contact, but differ in temperature 10° or 12°. They pursue their opposite courses for more than 1000 miles, and according as a vessel is placed in one or other of these currents will her progress be aided or retarded at the rate of from 40 to 50 miles a-day. The diagrams at the foot of the Map show—(1) The globe divided into two hemispheres—the northern, of which Britain is nearly the centre, containing the greatest amount of land; and the southern, the most central part of which is New Zealand, the greatest amount of water: (2) The thickness of the crust of the globe, as estimated from the highest known mountain and the deepest measured part of the sea, the difference being upwards of 14 miles.

c

## PLATE XII.—River Systems.

In this Map the surface of the globe is divided into great river-basins, showing by colours the different oceans, seas, and lakes into which the flowing waters of the different continents discharge themselves. Thus, in the Old World—Europe and Asia—the line of water-parting follows an irregular curve from Norway to Behring Strait. To the north of this more or less elevated ridge, all the waters flow northward to the Arctic Ocean, while to the south of it the waters of Scandinavia and part of Russia flow through the Baltic to the North Sea, and those of Eastern Asia flow to the North Pacific. Between this great line of water-parting in the north, and the rivers of Persia, India, and China in the south, the lake region of central Asia forms the "Basin of the Continental Streams," over the vast expanse of which all the rivers, including the Volga and many other large streams, discharge themselves into inland lakes, and do not directly reach the sea. South of the Basin of the Continental Streams, the rivers of Asia Minor, Beloochistan, India, Malaysia, the greater part of Australia, and Eastern Africa, flow to the Indian Ocean. The whole of Europe, west of a line drawn from the Caucasus to the North Cape, and in Africa the great basins of the Nile, Niger, Congo, and Orange rivers, occupying more than half its area, drain to the Atlantic. An almost riverless continental basin extends, in Africa, from Algeria to Lake Tchad.

In the New World—America—the north and south line of water-parting commences at the Arctic circle west of Hudson Bay, crosses the low ridge east of Great Slave Lake to the sources of the Saskatchewan River, and thence north-west to Behring Strait. All the rivers to the north of this flow to the Arctic Sea. The east and west line of water-parting, beginning at Behring Strait on the west, runs in the direction of the coast south-east to the sources of the Saskatchewan; thence skirting the eastern base of the Rocky Mountains, the mountains of Mexico and Central America, it enters South America, and follows the crest of the Andes, southward, to Patagonia. All eastward of this extensive line, the rivers flow east to the Atlantic Ocean; all west of it, to the Pacific. Within the double range of the Rocky Mountains in North America is the interior basin of the Great Salt Lake, 4000 to 5000 feet above the sea-level; and in South America, the basin of Lake Titicaca is 12,000 feet above the sea. In the diagram at the foot of the Plate, the principal rivers of the different continents are, for the sake of comparison, drawn to the same scale as that of the chart, and their lengths are stated in English miles. This Plate shows also the limit of permanently frozen ground in the northern hemisphere, which includes the greater portion of the Arctic basins of Asia and America, as well as the tracks of ships by the usual trade-routes.

## PLATES XIII., XIV.—Hydrographic Map of the British Isles.

The inland waters of the British Isles distribute themselves naturally, according to the cardinal points, into Northern, Eastern, Southern, and Western systems. The boundaries of these systems, the great water-partings of the islands, are shown on the map by strong red lines, while the minor divisions of these systems into river-basins are indicated by lighter lines of the same colour, the lengths and drainage areas of all rivers whose basin exceeds 400 square miles being given on the Map.*

* The lengths of the rivers and areas of the river-basins of England gives have been taken from the ' Plan of the Catchment Basins of the Rivers of England and Wales,' published at the Ordnance Survey Office in 1861. Those for Scotland and Ireland are the results of a careful series of original measurements from large-scale maps.

The elongated form of the island of Great Britain, in the direction from north to south, causes the line dividing the eastward from the westward flowing waters to be the most important in the island; and since the more elevated parts of the land are in the west, this line, following the highest ground, lies uniformly nearer to the west coast than the east. In Ireland the main water-parting is that which runs from north-west to south-west in a curve approaching the east coast in its centre, showing that, contrary to the case of Great Britain, the greater drainage of the land is to the westward. More than one-half of the area of Britain is drained to the eastward, and nearly one-half of the area of Ireland to the westward.

The following table gives the areas of the different systems of the British Isles in English square miles :—

| | Northern. | Southern. | Western. | Eastern. |
|---|---|---|---|---|
| England, . . . | ... | 7,305 | 21,255 | 29,759 |
| Scotland (Mainland), . | 1120 | 2,700 | 7,637 | 15,356 |
| Great Britain, . . . | 1120 | 7,306 | 31,592 | 45,115 |
| Ireland, . . . | 4800 | 6,800 | 15,750 | 5,150 |
| British Isles, . | 5920 | 14,106 | 47,342 | 50,265 |

From this table it will be seen that the Southern system of England is very small in comparison with its Western and Eastern systems, being only one-third of the former and one-fourth of the latter in area; and that the smallest system of all in the islands is that of the north of Scotland, whose extent is only a fortieth part of that of the greatest, the Eastern system of Britain.

If Scotland be considered by itself separately from England, it may be said to have a southern drainage system of 2700 square miles in extent to the Irish Sea; but if taken in connection with England as forming with it the island of Great Britain, this area must be looked upon as a part of the Western drainage system of the island. The Northern, Southern, and Eastern systems of Ireland are of nearly equal size, and the three taken together form an area not much greater than that of the Western system.

The Eastern system of Great Britain, included within the water-parting line which runs from Duncansby Head, on the north-east, to near Dover, on the south-east, and occupying more than one-half of the area of the island, embraces most of the larger rivers of Britain, and discharges its waters to the North Sea. Eleven of the river-basins of this system have an area of upwards of 1000 square miles, and are those of the Spey, Tay, and Tweed, in North Britain; and of the Tyne, Ouse, Humber, Trent, Witham, Nen, Great Ouse, and Thames in South Britain—the last named, the Thames, being the greatest river of the British Isles, and draining an area of 5162 square miles.

The Western system of Great Britain, stretching from Cape Wrath to Land's End, and draining to the Atlantic and the Irish Sea, has but four river-basins of greater extent than 1000 square miles, these being the Clyde in Scotland, and the Mersey, Wye, and Severn in England. The northern part of this system has no rivers properly so called, but is drained by numerous streams, which rise with rain and fall with dry weather; but in the south of it, the Severn, with its basin of 4437 square miles, is the third river of the British Isles.

The Southern system of Britain is a narrow strip of land extending between Dover and

Land's End, draining to the English Channel, and having no great rivers—the largest basin, that of the Avon, being only 666 square miles in extent.

The Northern system in Britain is one of streams alone; the largest, the Naver, having a basin of only 200 square miles.

The Southern, Western, and Northern systems of Ireland send their waters to the Atlantic. The Eastern system drains to the Irish Sea.

The Western system of Ireland, like the Eastern of Britain, embraces about one-half of the area of the island. The water-parting bounding it runs from Bloody Foreland in the north-west to Mizen Head in the south-west, including three river-basins of upwards of 1000 square miles in area—those of the Erne, Galway, and Shannon; the last the largest river of Ireland, and the second of the British Isles, draining an area of 4590 square miles. The water-parting of the Southern system of Irish rivers runs inland to near the centre of Ireland, round the head-waters of the Barrow, its largest river, and meets the coast at Carnsore Point on the east, and Mizen Head on the west. Besides the Barrow, the Blackwater is the only other river of this system with a basin of more than 1000 square miles.

The Eastern system of Ireland extends in a narrow strip from Fair Head on the north to Carnsore Point in the south. Its only river-basin of above 1000 square miles in area is that of the Boyne.

Almost the entire area of the Northern system of Ireland which lies between Fair Head and Bloody Foreland is occupied by the rivers Bann and Foyle, each of which drains an area of upwards of 1000 square miles.

In the order of the extent of their basins, the rivers which drain an area of upwards of 1000 square miles stand as follows :—

| | Sq. miles. | | | Sq. miles. |
|---|---|---|---|---|
| 1. Thames, | 5162 | | 13. Wye, | 1655 |
| 2. Shannon, | 4590 | | 14. Blackwater, | 1300 |
| 3. Severn, | 4437 | | 15. Spey, | 1245 |
| 4. Ouse, | 4207 | | 16. Humber, | 1239 |
| 5. Trent, | 3972 | | 17. Galway, | 1172 |
| 6. Barrow, | 3517 | | 18. Clyde, | 1145 |
| 7. Great Ouse, | 2634 | | 19. Foyle, | 1090 |
| 8. Bann, | 2265 | | 20. Nen, | 1055 |
| 9. Tay, | 2090 | | 21. Tyne, | 1053 |
| 10. Tweed, | 1990 | | 22. Witham, | 1052 |
| 11. Mersey, | 1706 | | 23. Boyne, | 1046 |
| 12. Erne, | 1660 | | | |

The contour lines on the Map are of 250, 500, 750, and 1000 feet, and have been worked out chiefly from the books of the Ordnance Spirit-Levelling Survey. Though they can only be said to be an approximation to the truth, yet they give a much more accurate representation of the elevation of the land than can be obtained by any system of hill-shading.

From these lines an estimate may be made of the elevation of the source and the average rapidity of fall of any river, by observing its passage from one height to another. If the contours approach closely to one another in a river-course, its fall is most probably rapid, and the river unfit for purposes of navigation; but if, on the contrary, the lines be far apart, and at equal distances from each other, we may predicate that the river has a gentle descent to the sea.

The blue tints representing the rainfall have been obtained by taking the mean of the rainfall, as given in Mr Symons's reports* for the five years from 1863 to 1867 inclusive, of nearly 500 stations in the British Isles. The figures thus found were set down on large-scale maps, and lines were drawn through the places which have the same average rainfall, whilst keeping

* 'British Rainfall.' Compiled by G. J. Symons, F.M.S.

in view the general fact that a greater elevation gives a greater rainfall. From these tints it is evident that the western parts of the islands receive the greatest amount of rain. This may be due to two causes—first, to the prevalence of rain-bringing south-west winds ; and second, and perhaps chiefly, to this, that the western parts of the islands are generally of greater elevation than the eastern.

The part of the British Islands which shows the least rainfall is that strip of the east coast which lies between the mouth of the Humber and the Wash, the rainfall there being under 20 inches. A small portion of the north coast of Elginshire has also, perhaps, this minimum of rainfall.

Almost the whole of eastern England has a rainfall of from 20 to 25 inches—the Downs in the south and the Wealds in the north being the exceptions ; but only some small patches of the east coast of Scotland have this rainfall, and the fall in all parts of Ireland seems to be greater than this.

The line of 30 inches of rainfall runs as far back as the Welsh boundary in Mid-England, but is carried eastwards again by the Pennine chain and the Cheviots, and includes only the east coast of Scotland, though brought considerably inland by the valley of the Tweed, the low land between the rivers Forth and Clyde, and Strathmore of the Tay Basin.

The lowest rainfall in Ireland, as far as has yet been ascertained, is from 25 to 30 inches on the lower parts of the south-east and east coasts. The central plain of Ireland lies entirely within the lines of 30 to 35 inches of fall. The lines of 35 to 45 inches and above are confined to the higher parts of west and south-west England, and the Pennine chain in the north, but include the great mass of Scotland and the whole of Ireland, excepting the central plain and the east coast. The observing stations on the west coasts and in the hilly regions are not sufficiently numerous to enable the lines of greater rainfall to be drawn on the Map ; but it is most probable that these lines would be found to follow the same law which holds good in the other parts of the islands, of increase with greater elevation and proximity to the western coasts.

The maximum observed rainfall seems to be that noted at Seathwaite, in the Cumberland mountains, which, in the mean of the five years from 1863 to 1867, has a fall of 140 inches ; but many of the higher parts of the west of England, Scotland, and Ireland have doubtless a rainfall of more than 100 inches.

### PLATE XV.—Rain and Snow.

Rain is distributed very unequally over the globe. In general it is most abundant in those regions where evaporation is carried on most rapidly ; but there are striking exceptions to this rule, for in many places, even near the tropics, it seldom or never rains. These rainless districts comprise vast regions, almost devoid of vegetation. On the contrary, there are regions where rain falls almost incessantly, and where, consequently, vegetation is rank and abundant. The zone of greatest precipitation is situated on the north of the Equator, and corresponds with the zone of "the variables" in Plate XVII. Within this belt, on the west coast of Africa, the average annual quantity is 189 inches. Rain is in general most abundant near the Equator, where the temperature is highest ; and the quantity decreases, irregularly, with the decrease of heat, as we approach the polar regions. The average annual fall within the tropics is 95 inches, and within the temperate zone, 34 inches. This is shown graphically in the diagram ("Increase of Rain with Heat") at the foot of the Plate. The diagram of "Decrease of Rain with Distance from Coasts," illustrates a general law occasioned by the greater amount of vapour that arises from the sea than from the land. In the temperate zone of both hemispheres, the western coasts are proportionally more moist than the eastern, because they are exposed to the prevailing westerly currents of air, which, passing over the ocean, are highly impregnated with moisture. Within the tropics, the eastern coasts, especially in America, are more moist than the western, from their exposure to the trade-winds. In certain districts more rain falls in one season than another : these are distinguished on the Map as the summer, winter, and autumn rains.

Snow never falls within or near the tropical regions at the level of the sea. The red lines on the Map show that the deposition of moisture in this form is, in the northern hemisphere, limited to the north of India, the Mediterranean, and the Gulf of Mexico; while in the southern hemisphere it does not approach nearer the Equator than the south of Australia, the Cape Colony, and Patagonia. But in tropical and sub-tropical regions snow falls on mountains where, owing to elevation above the sea, the temperature is reduced to the freezing point ; as in the Himalayas in Asia, Mounts Kenia and Kilimandjaro in Eastern Africa, and the Andes of South America. The fall of snow increases with the decrease of temperature, in proceeding from the equatorial towards the polar regions. In Europe this decrease occurs in the following order :— Rome has 1¼ snowy days in each winter, Venice 5⅓, Milan 10, Paris 12, Carlsruhe 26, Copenhagen 30, and St Petersburg 171. The actual elevation of the snow-line—the line of perennial congelation—or the lowest point on a mountain at which snow is never entirely melted, in the different zones, is shown in the diagram on the left of the map, and on an enlarged scale in Plates XVIII. and XIX.

### PLATE XVI.—Climatology *.—Isothermal Lines.

The distribution of heat over the globe is here rendered perceptible to the eye by lines drawn through all places having the same mean annual temperature. These are called *isothermals*, and their deviation from parallelism with the Equator shows that latitude, or distance from the Equator, alone, gives little indication of the temperature of any particular place. The mean annual temperature of the more important places on the globe is stated in figures on the larger Map. The Map of Central Europe shows by red and blue figures the summer and winter temperatures of the principal meteorological stations, while that of the British Islands presents the three elements of summer, winter, and mean annual temperatures. If these places were connected by lines, the first would be called *isothèral*, the second *isocheimönal*, and the third *isothermal* lines. July is, on an average, the hottest, and January the coldest, month of the year ; the mean annual temperature occurs in April and October in the north temperate zone. On each side of the meridian of 30° west, the extreme temperature of summer and winter (July and January), for the latitude, is stated in red figures for summer, and blue for winter. Places where the temperature of the latitude agrees with the isothermal line are said to have a normal temperature ; those places where the isotherm is lower being relatively *colder*, and those where higher relatively *warmer*, than the normal, are said to have an abnormal temperature. A remarkable example of the latter occurs on the west coast of Europe, where the warm water of the Gulf Stream has the effect of carrying the isothermals many degrees north of their normal position. A climate is called an *insular* or *sea* climate where the difference of mean temperature is very small, or where the winter is milder and the summer cooler than the average ; and a *continental* climate where the difference of mean temperature is very great, or where the win-

* See Buchan's 'Handy Book of Meteorology,' 2d edition, Edinburgh, 1868, where the whole subject of Climatology is treated of in an interesting and lucid manner.

ter is very cold and the summer very hot. Thus the differences between the summer and winter temperature of a country increase with distance from the sea. Europe has a true insular climate—a mild winter and a cool summer. Northern and Central Asia have a true continental climate—a cold winter and a hot summer; while North America has more of a continental climate in winter, and a sea climate in summer. The hottest locality on the globe is in northern Central Africa (shaded red on the Map), where the temperature of July is 90° Fahr.; and the coldest in Siberia (shaded blue on the Map), where the temperature of January is 40° below the freezing point of Fahr. scale. In the old continent the cold comes from the north-east, and in the new from the north-west. The cold region of Siberia has no corresponding region of equal cold in America. If the globe be divided at the meridian of 20° W., we find that the eastern portion, which has the largest mass of land, is colder than the western, and that the difference diminishes as we approach the Equator. The temperature of the whole globe increases 8° Fahr. from January to July; a mean between these months gives, as the mean temperature of the globe, 58° Fahr. The mean temperature for the northern hemisphere is 60° Fahr., and the mean for the southern hemisphere 56° Fahr. The great quantity of rain which falls in the northern hemisphere is probably one cause of its higher temperature, while, in the southern hemisphere, the influence of the sun's rays is expended to a great degree in the melting of masses of ice, or in evaporation from the surface of the ocean.

### PLATE XVII., Winds and Storms,

presents a graphic representation of the principal phenomena of the winds over the globe. Near the centre is the great equatorial belt of calms and variable winds and storms, corresponding to the zone of constant rain in Plate XV., and on the north and south of the trade-wind region, the narrower belts of the calms of Cancer and Capricorn. The region of the *Trade* or passage winds, called *Vents alisés* by the French, extends to about 30° on each side of the Equator, and beyond these are the regions of the south-westerly and north-westerly currents of air. The Indian Ocean is the chief region of the *Monsoon winds* (from the Arabic "Mausim," an epoch or season), included within which is the district of the *Typhoons*, or storms peculiar to the China Sea. But monsoons, or monsoon-like winds, prevail to a limited extent in various parts of the globe, as on the shores of Brazil, Chile, Peru, Mexico, and the west and north coasts of Africa, indicated by the red colour on the chart. The Red Sea and the Persian Gulf, not being affected by monsoon winds, are left white.

The diagrams at the foot of the Plate are devoted to an explanation of the *Hurricanes* of the West Indies and the Indian Ocean, and of the varying limits of the trade-winds in the Atlantic, according to the different seasons of the year. The hurricanes of the Indian Ocean and of the West Indies, and the typhoons of the China Sea, appear to be subjected to peculiar and fixed laws for each hemisphere, both as regards their movements of translation and their rotatory movements. These tempests originate between the Equator and the tropics during winter, when the regularity of the trade-winds is interrupted, or during the change of the monsoons. They obey a double movement—the one of translation or progress, and the other of a gyratory or rotatory kind. On the north of the Equator the rotatory movement is from *right* to *left*, while on the south of the Equator it is reversed, or passes from *left* to *right*, as shown on the Plate. The movement of translation takes the form of a parabolic curve, the summit of which is towards the west. In the northern hemisphere this curve forms a tangent to the meridian about lat. 30°, and in the southern hemisphere about lat. 26°, corresponding with the northern and southern limits of the trade-winds. In the Indian Ocean, these storms uniformly come from the eastward, and travel to the westward and southward, as shown in the chart of the Rodriguez hurricane. The West Indian hurricanes commence near the Leeward Islands, travel to the W.N.W., and then round the shores of the Gulf of Mexico, following the course of the Gulf stream, and are lost in the Atlantic between the Bermudas and Halifax. The chart in the left corner of the Plate shows the dates of occurrence, and the courses, of some of the most important of these storms. The rate of progress of a hurricane varies in different parts of its course; that of the Rodriguez storm has been calculated at from 220 to 230 miles a-day at first, diminishing, as it approaches the tropics, to about 50 miles a-day.

The districts subject to local storms are marked in the Map as the *Sirocco* in the Levant, the *Simoom* in Arabia, the *Khamsin* in north-east and the *Harmattan* in south-west Africa, the *Pamperos* in South America, the *Northers* in Mexico, and the *Hot Winds* in Australia.

## PLATE XVIII.—Vegetable Life.

The plants of any particular region are the exponents of its climate: certain plants will grow spontaneously only within certain districts or zones, the boundaries of which are dependent on the amount of heat and moisture which such zones receive in the course of the year. Hence we find that trees, grain, and shrubs range themselves on the globe according to lines of equal mean summer and equal winter temperature: the lines of summer temperature, for example, indicate precisely the limits of the possible cultivation of annual plants. Nor is the knowledge of the capabilities of a country for producing plants less important with reference to its population. Comparing Naples with Norway, for example, we find that the effect of climate is such as to render the harvest five times more productive in the former than in the latter place: while, in consequence, the population is twenty-five times more dense, in proportion to its area, in Naples than in Norway.

The object of this Map is, by dividing the whole earth according to its peculiar flora into certain climates of vegetation, to present at one view the distribution of the most useful and valuable wild and cultivated plants. Each of the plant-climates is characterised by certain trees, grains, and fruits, the number and variety of which increase as we approach the Equator, and decrease towards the polar regions, where their only representatives are a few mosses and grasses. Rice, which supports the greatest number of the human family, is chiefly confined to the tropical regions of Asia and America. Wheat requires a mean annual temperature of 37° or 39° Fahr., and a mean summer temperature of 56° to 58° at least. Maize extends to lat. 40° N. and S. in America, and to lat. 50° or 52° N. in Europe; barley to lat. 70° N. in Norway; rye to lat. 67°, and oats to lat. 65° N. Since heat decreases as we ascend into the atmosphere, so mountains situated between the tropics, the summits of which rise above the snow-line, represent the vegetable zones of the whole earth, rising one above the other *vertically*, in the same order as they observe in a *horizontal* direction from the Equator to the Poles, on plains. This is explained by the colours on the diagram at the bottom of the Plate, on which also is represented the actual ascertained limits of the snow-line in different latitudes. In the Himalaya this will be observed to be 19,000 feet on the north, while it is only 15,500 feet on the south side—an anomaly which has been considered due to the radiation of heat from the high land of Tibet, but which is more probably owing to the greater quantity of snow which falls on the southern than on the northern slopes of the mountains.

## PLATE XIX.—Animal Life.

Animals, like plants, are adapted to special climatic conditions; like them, also, they are subjected to invariable laws. Each zone, or region of climate, is occupied by some species of animals peculiar to itself, beyond the limits of which they will not range if left to their natural freedom. A group of animals, embracing all the species both terrestrial and aquatic, inhabiting any particular region, is called a *Fauna*, in the same way as the plants of any particular country constitute its *Flora*. In this Map the faunas of the two hemispheres are distributed in three principal divisions—namely, the Tropical, Temperate, and Arctic faunas, each of which is characterised by peculiar species of animals. These, like plants, attain their highest development within the torrid zone, which is distinguished not only by the great variety, but by the size, strength, and beauty of its animal formations. The mean density of species inhabiting the different zones of the earth strikingly proves this. The proportion is estimated thus: in the tropical region, 26 species; in the north-temperate, 9; and in the arctic, only 7 of the whole. In proportion as they recede from the Equator, so the covering of animals loses its brilliancy of colour, until, in the polar regions, all animals, whether marine or terrestrial, assume a nearly uniform and sombre hue. From a comparison of this with the preceding Map, it will appear that the nature of their food has an important bearing on the distribution and the grouping of animals, those of the herbivorous kind being more or less limited to particular vegetable zones, while, since the food of carnivorous animals is everywhere present, their range is, in this respect, much less confined, the absence of animal life itself being the only restriction.

The diagram at the foot of the Plate represents the distribution of animals in a vertical direction, showing, as in the case of plants, that those nearest the level of the sea correspond with those of the torrid zone; while, on approaching the snow-line near the tops of mountains, the animals are similar to those of the arctic zone.

## PLATE XX.—Races of Man.

The geographical distribution of man is different from that of all other organic beings: his constitution renders him more cosmopolite, while, from the superior structure of his physical frame, as well as his mental endowments, he is less subject to the influence of external circumstances than any of the inferior animals: hence he is found, although not in equally favourable circumstances, in every locality over the globe, under every climate, and at every degree of altitude to which organic life extends. All the various races of man existing on the globe have originated from one species, but distinct races have been known to exist since the earliest dawn of history and tradition. The principal map exhibits the distribution of the leading races of man, as existing at the present time, the localities of which are explained by the reference figures and colours.

The Ethnographic Map of Europe presents a more minute subdivision of these races in this quarter of the world. It distinguishes by colours three great varieties of the Caucasian division—the Teutonic, the Celtic, and the Sclavonian. The sub-varieties have different shades of these fundamental colours; and wherever there has been a crossing of these varieties, it is indicated by a mixed tint. The diagram in the right corner of the Plate explains, by means of the scale of feet, the height above the level of the sea of inhabited places in different countries.

D

Coloured lines and a scale placed near the meridian of 20° W. on the upper Map, indicate the prevailing kinds of food used by the inhabitants of the different zones. Thus within the tropics the principal food of the human family is derived from the vegetable kingdom; beyond the tropics, north and south, a mixed animal and vegetable diet is used; while north of the Arctic circle the food of man is derived entirely from the animal kingdom. A faint line passing through Greenland and the north of Asia points out the northern limit of permanent habitation; the corresponding line in the southern hemisphere passes south of Cape Horn and Tasmania.

FRONTISPIECE

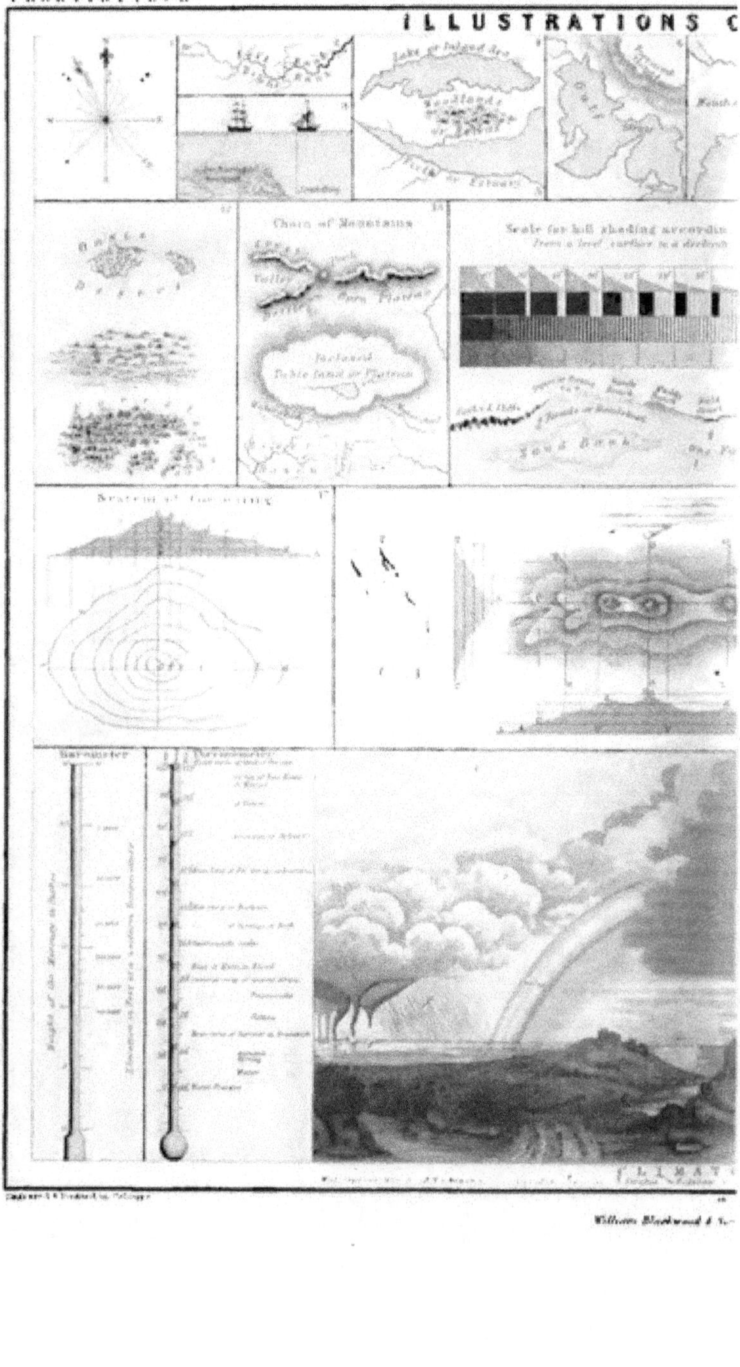


Engraved by W. & A.K. Johnston, Edinburgh

William Blackwood & So

PLATE I

# F CHARTOGRAPHY

ILLUSTRATIONS OF THE ACTION OF RAIN AND STREAMS

Diagram section of a Waterfall

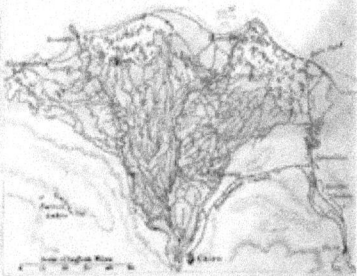

Serpentine Course of a River on flat ground

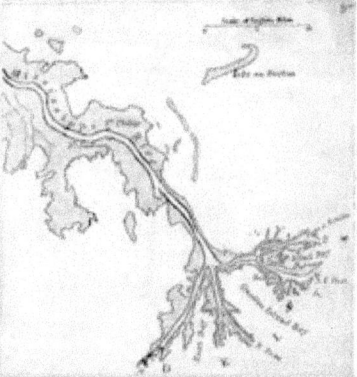

William Blackwood & Son

ILLUSTRATIONS OF THE ACTION OF ICE AND SNOW

## ILLUSTRATIONS OF SEA ACTION.

SECTIONS ACROSS THE BASIN OF THE ATLANTIC OCEAN

a. From Trinity Bay, Newfoundland, to Valentia, Ireland, on the line of the Telegraph Cables. Length 1950 miles.

b. From Cape Race, Newfoundland, to C.º S.º Roque, Brazil. Length 3500 miles.

c. From C.º S.º Roque, Brazil, to C. Palmas, W. Africa. Length 1800 miles.

William Blackwood & Sons.

ILLUSTRATIONS OF VOLCANIC ACTION AND MOVEMENTS OF THE EARTH'S CRUST.

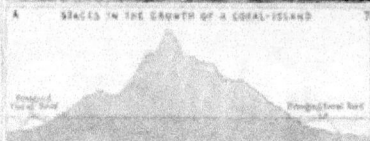

STAGES IN THE GROWTH OF A CORAL-ISLAND

Atoll or Coral Island

PROFILE OF A HILL-SLOPE SHOWING RAISED SEA-BEACHES

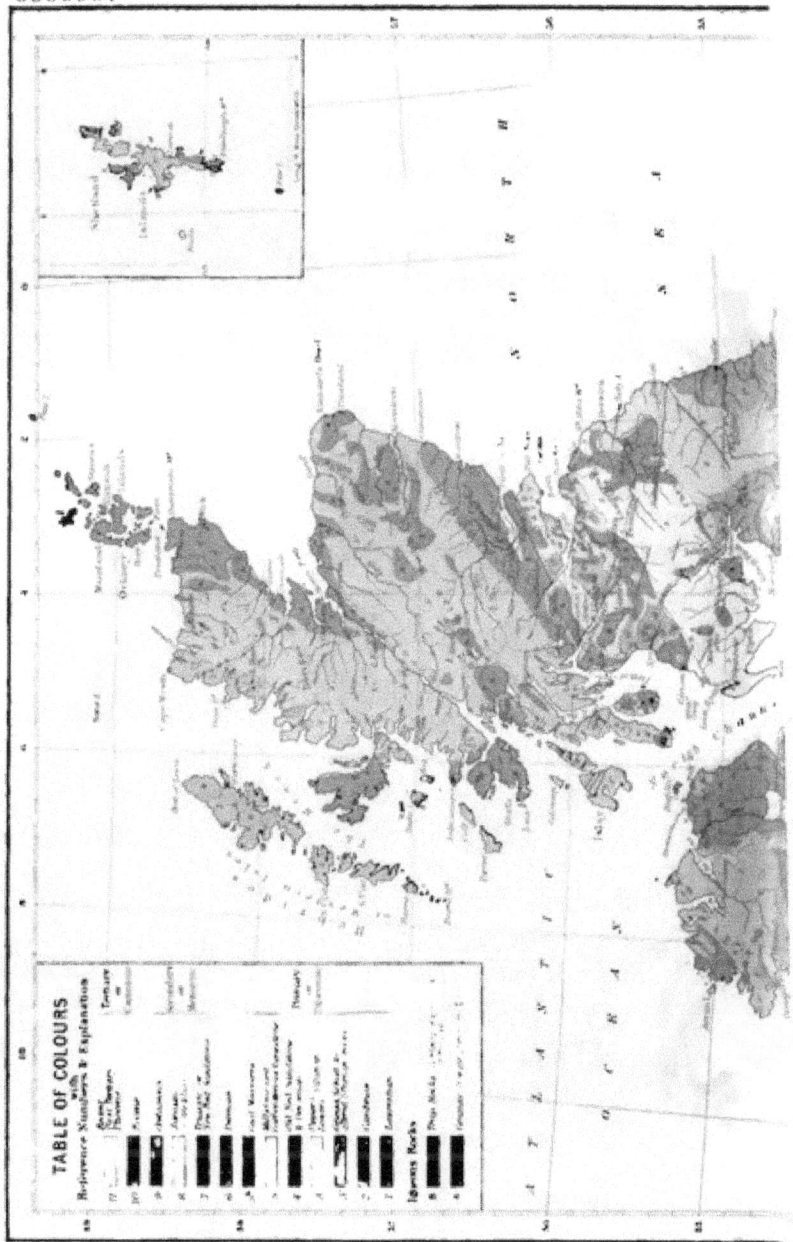

PLATE 4.

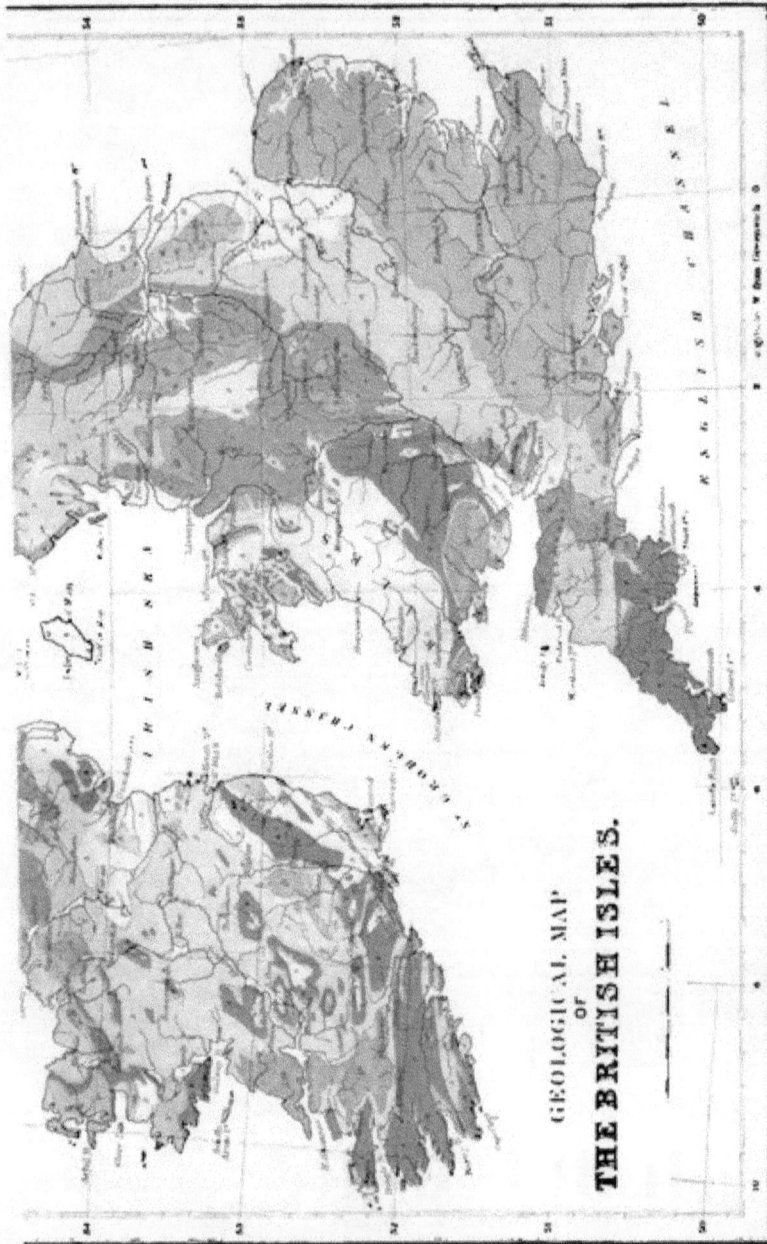

GEOLOGICAL MAP
OF
THE BRITISH ISLES.

ENGLISH CHANNEL

IRISH SEA

ST GEORGES CHANNEL

William Blackwood & Sons, Edinburgh & London.

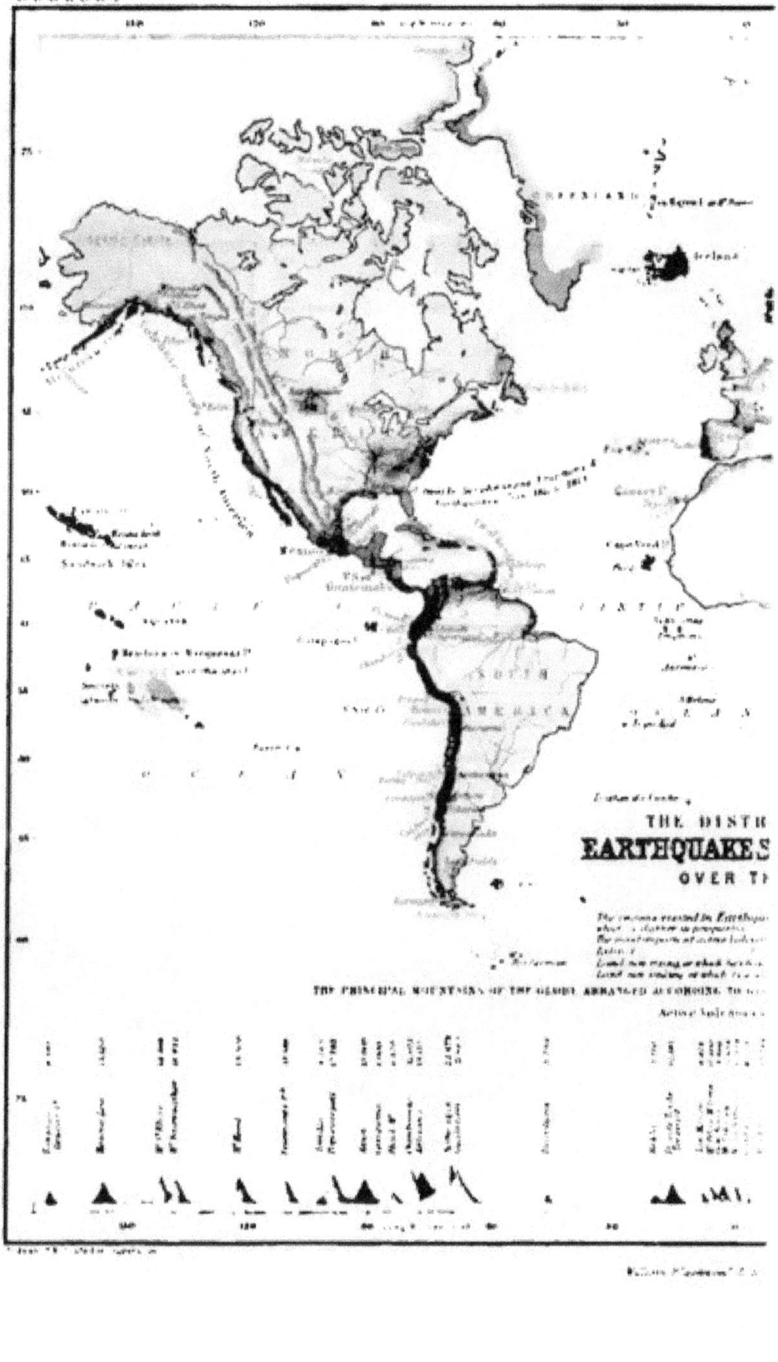
Iceland
Sandwich Isles
Mexico
SOUTH
AMERICA
THE DISTR
EARTHQUAKES
OVER TH
THE PRINCIPAL MOUNTAINS OF THE GLOBE ARRANGED ACCORDING TO

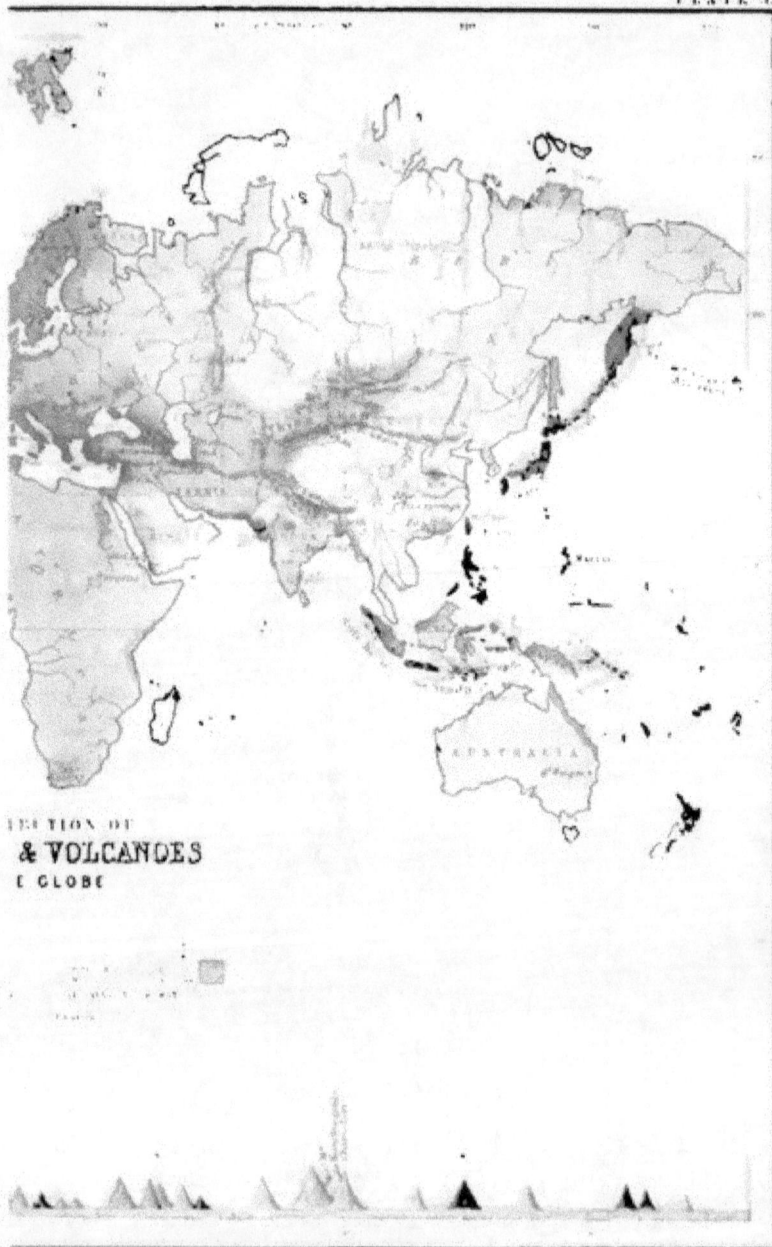

PLATE 5.

& VOLCANOES
E GLOBE

William Blackwood & Co

PLATE 6.

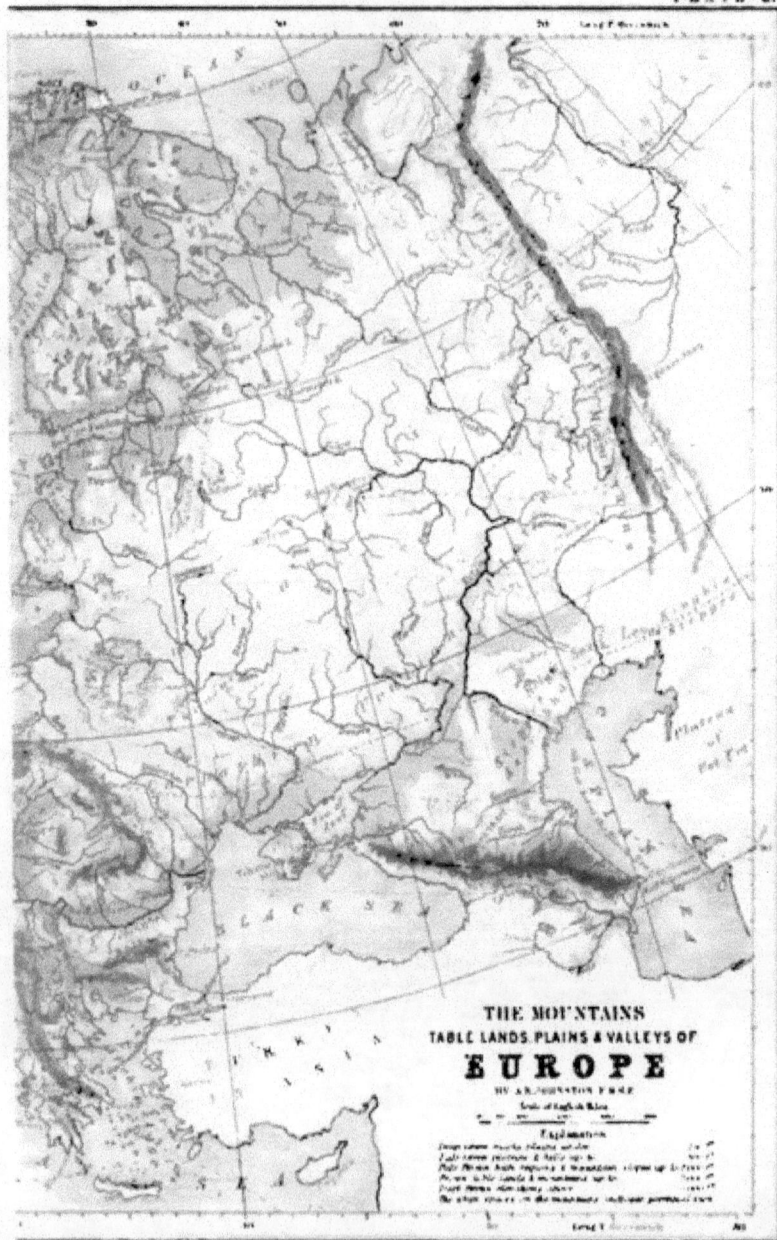

THE MOUNTAINS
TABLE LANDS, PLAINS & VALLEYS OF
EUROPE

BY A.K. JOHNSTON F.R.S.E

Scale of English Miles

Explanation

Edinburgh & London.

THE MOUNTAINS,
TABLE LANDS, PLAINS & VALLEYS OF
ASIA

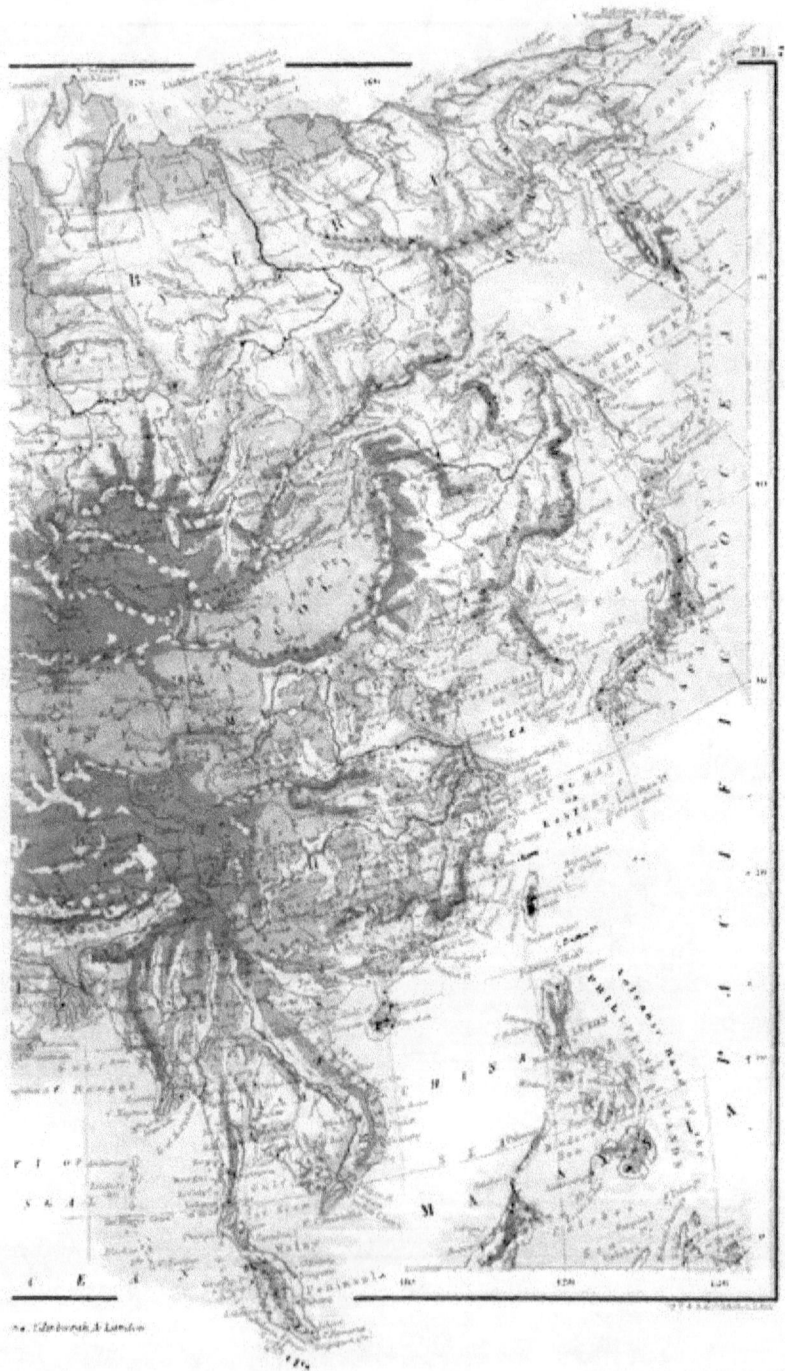

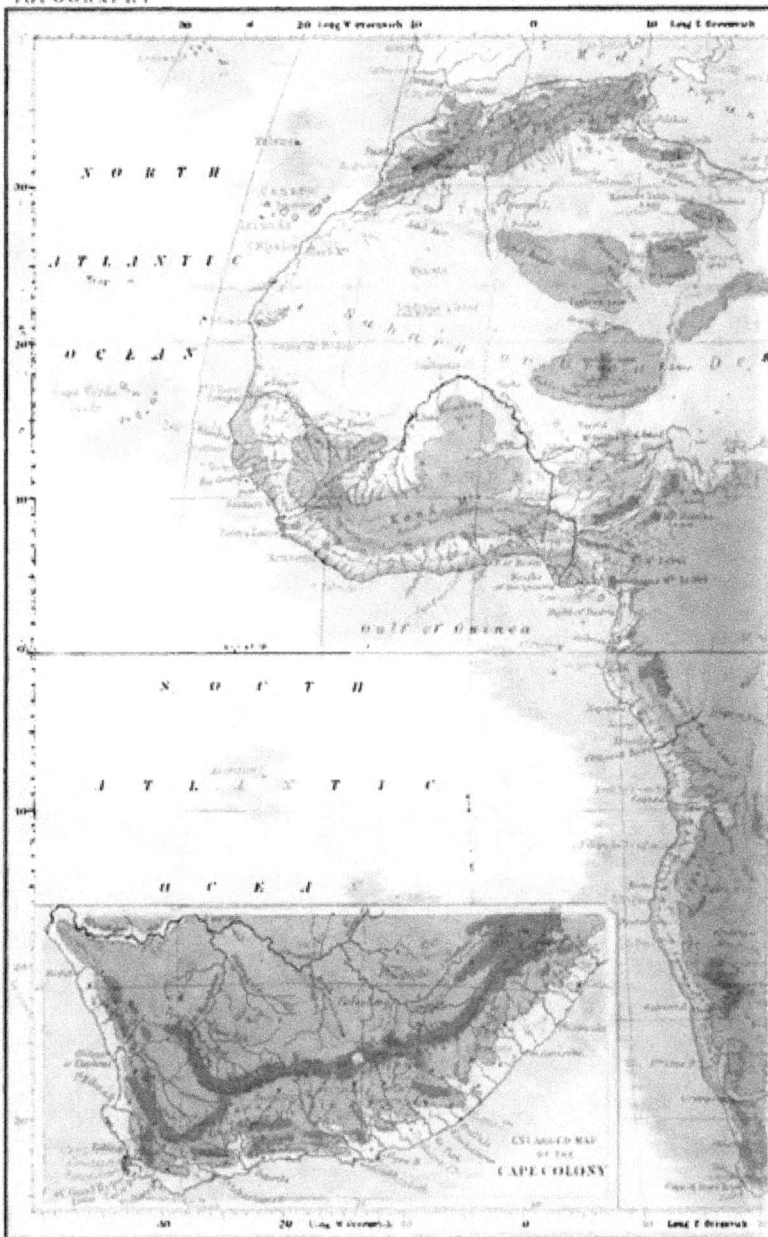

ENLARGED MAP
OF THE
CAPE COLONY

William Blackwood & Sons.

PLATE 8

THE MOUNTAINS
TABLE LANDS PLAINS & VALLEYS OF
AFRICA

PLATE 9

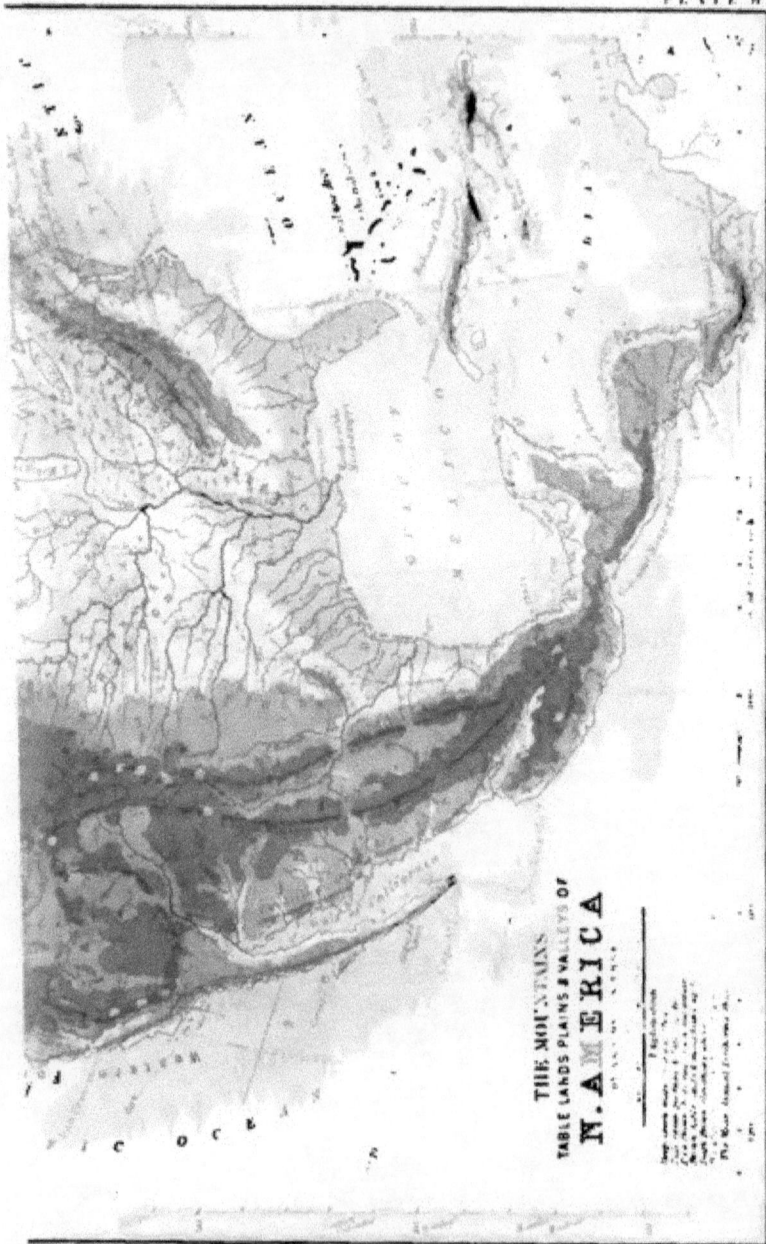

THE MOUNTAINS
TABLE LANDS PLAINS & VALLEYS OF
N. AMERICA

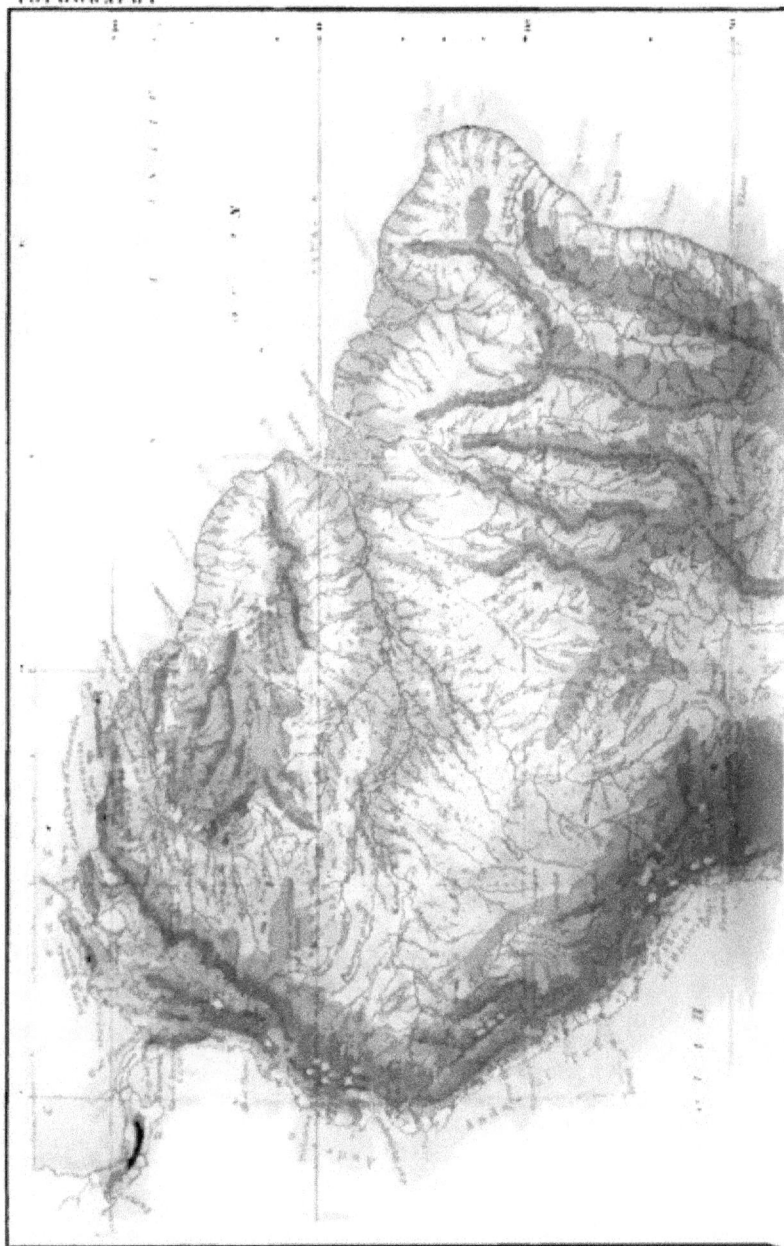

PLATE 10.

THE MOUNTAINS
TABLE LANDS PLAINS & VALLEYS OF
SOUTH AMERICA

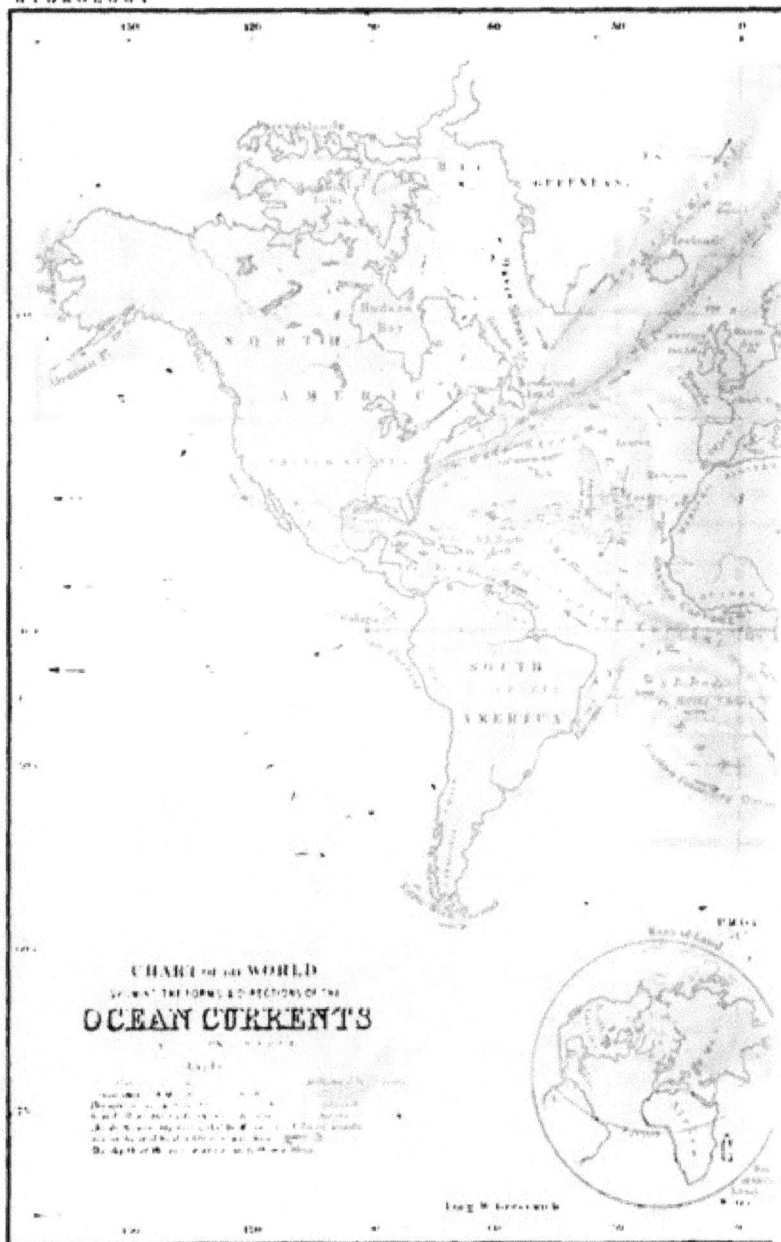

CHART OF THE WORLD
SHOWING THE FORMS & DIRECTIONS OF THE
OCEAN CURRENTS

PLATE II

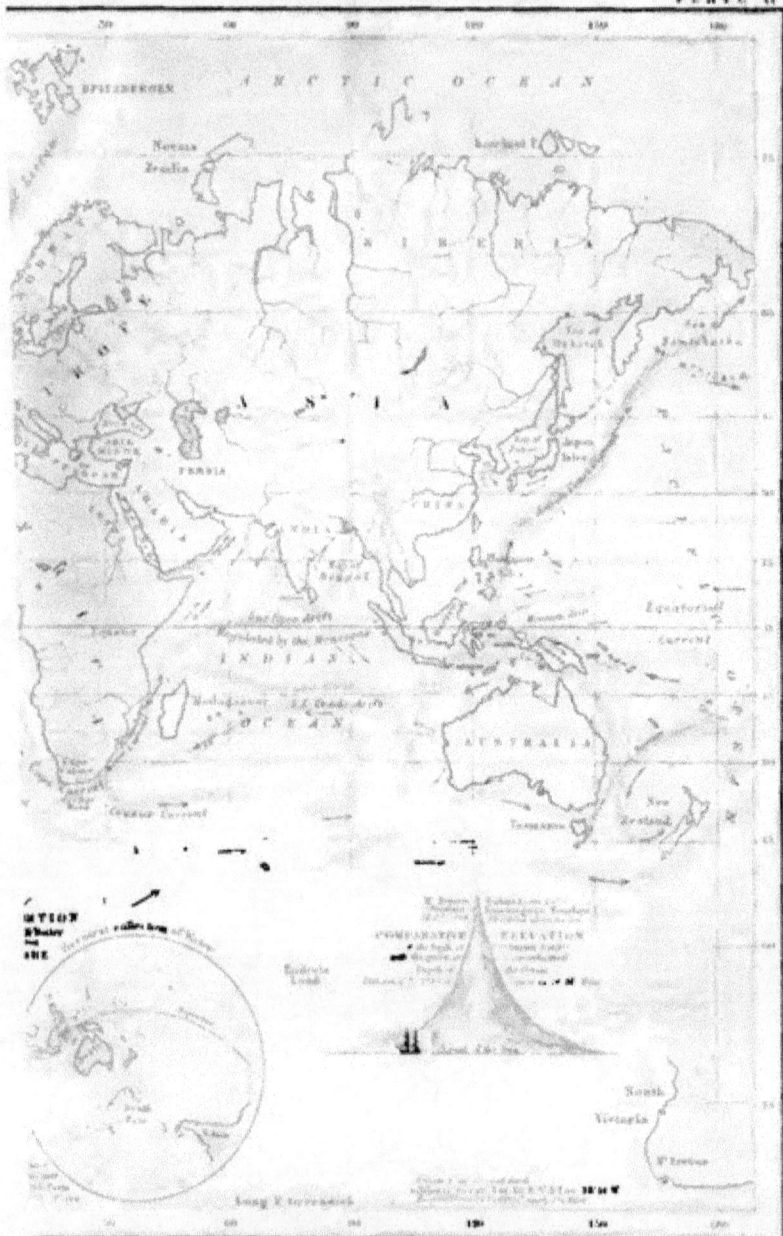

Comparative Lengths of the Principal Rivers

William Blackwood & Son

PLATE 12

ARCTIC OCEAN

NORTH

PACIFIC

OCEAN

INDIAN

OCEAN

SOUTH

PACIFIC

OCEAN

SYSTEMS
WORLD
VER BASINS
S. LAKES, &c

On a scale corresponding with that of the Map.

Europe Africa

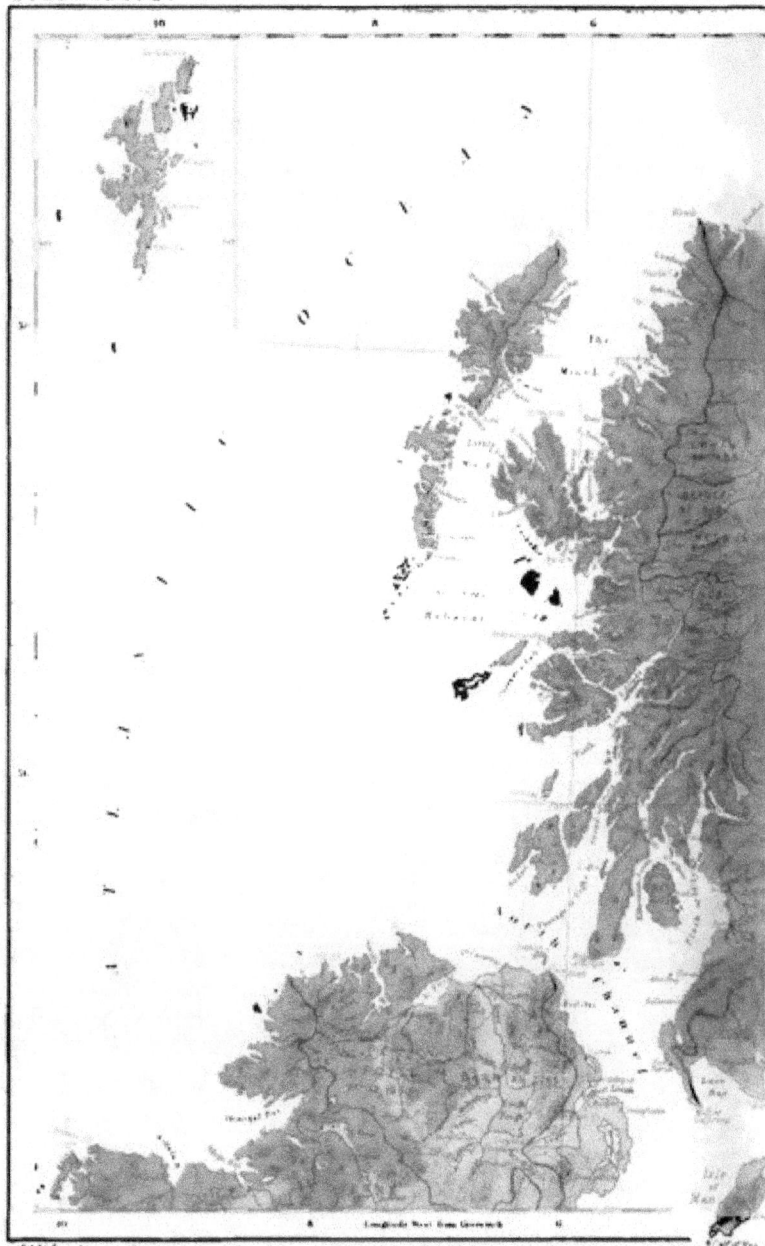

William Blackwood

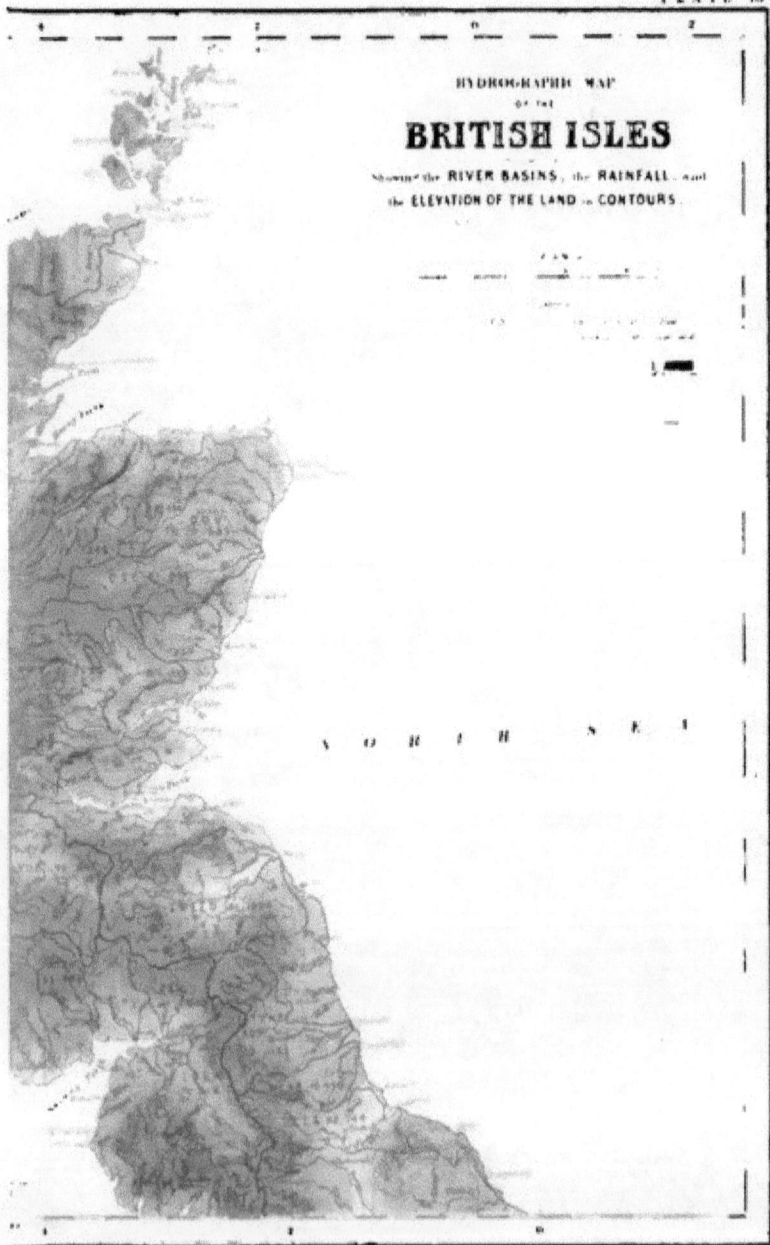

PLATE 13

HYDROGRAPHIC MAP
OF THE

# BRITISH ISLES

Showing the RIVER BASINS, the RAINFALL, and
the ELEVATION OF THE LAND in CONTOURS.

NORTH SEA

Edinburgh & London

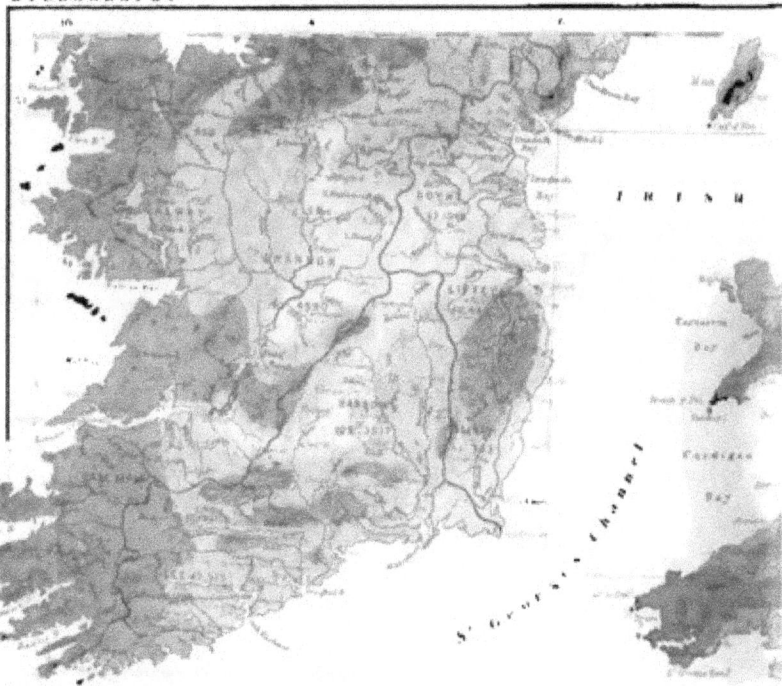

HYDROGRAPHIC MAP
OF THE
# BRITISH ISLES
Showing the RIVER BASINS, the RAINFALL, and
the ELEVATION OF THE LAND in CONTOURS.

Scale of English Miles

Explanations

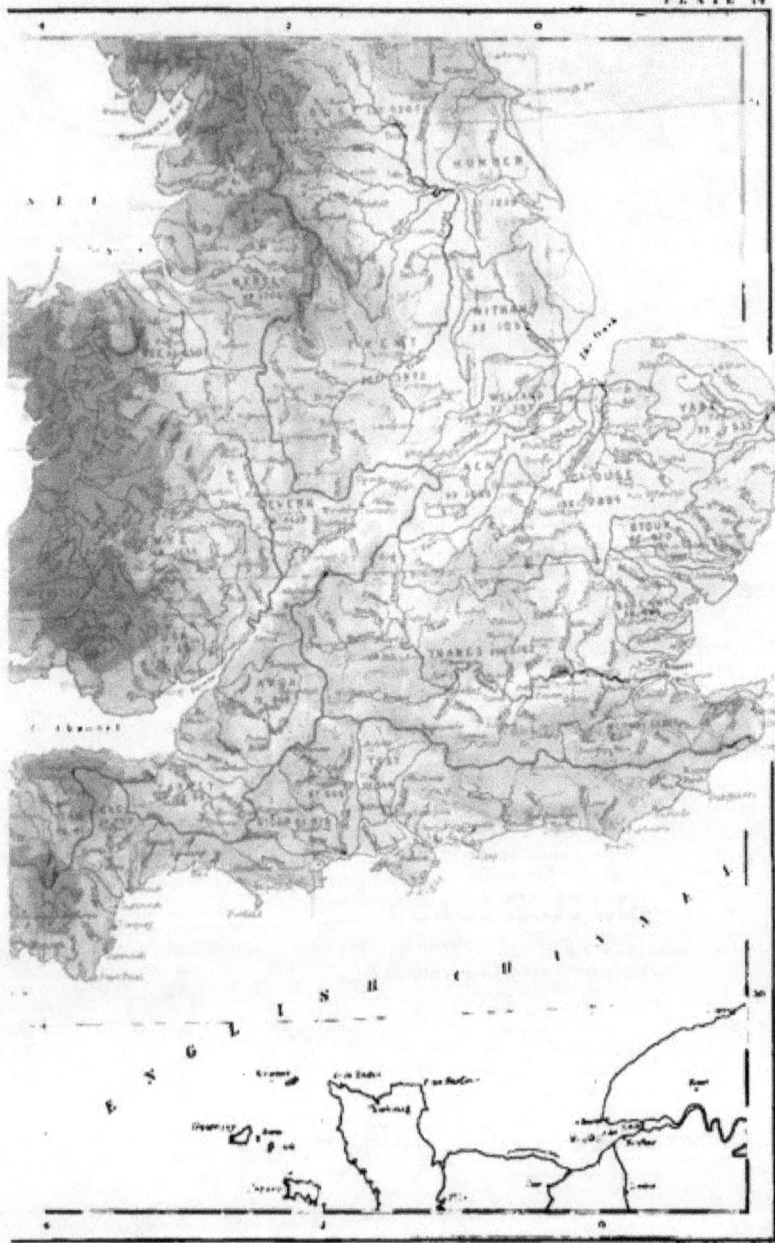

PLATE 14

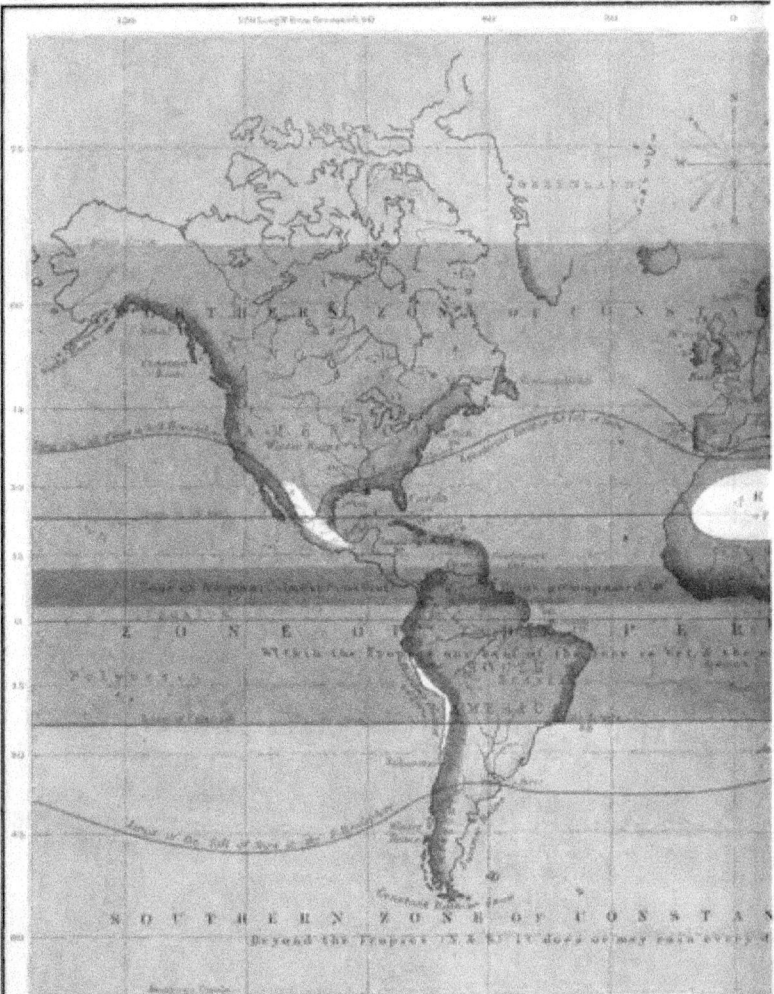

PLATE 15

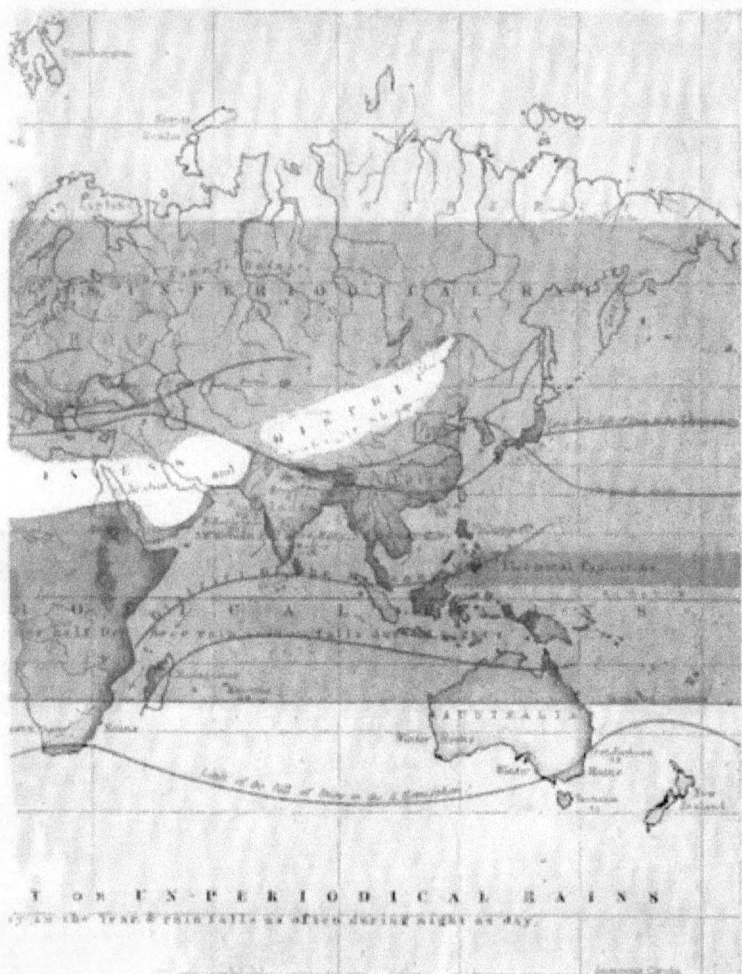

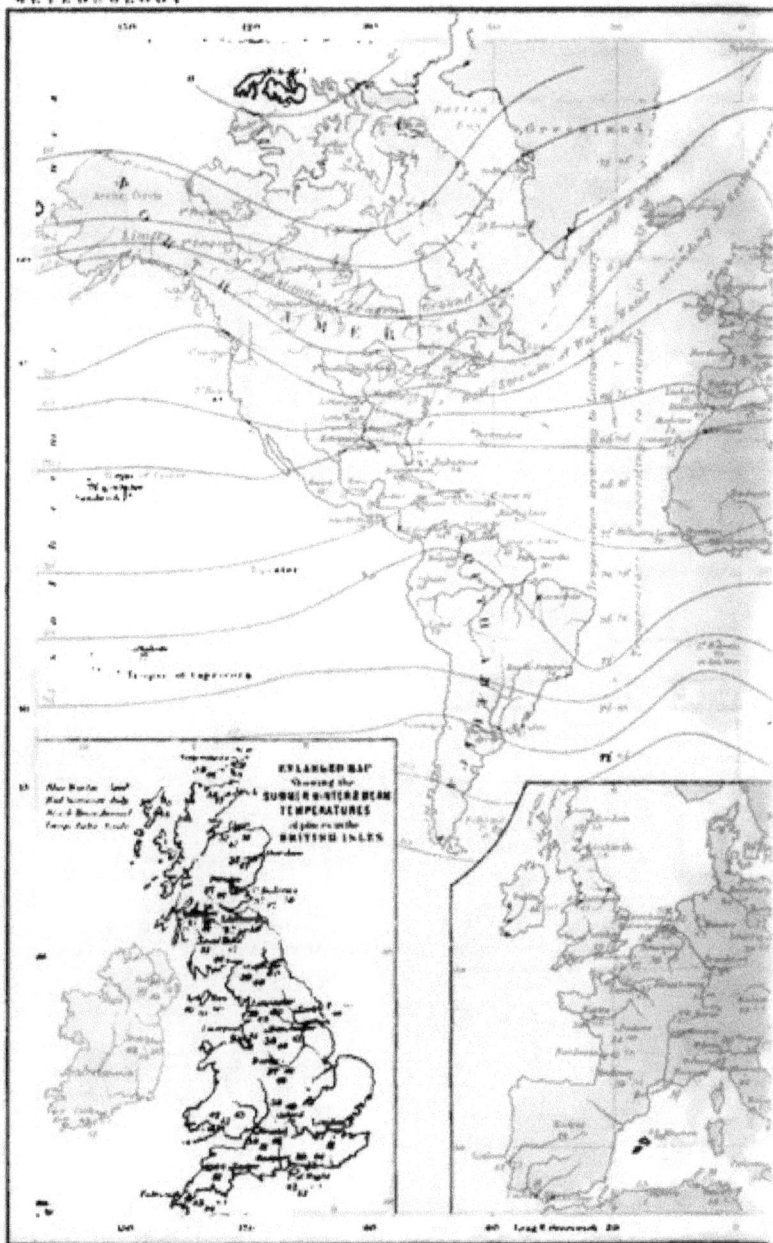

ENLARGED MAP
Showing the
SUMMER & WINTER MEAN
TEMPERATURES
of places on the
BRITISH ISLES

PLATE 16

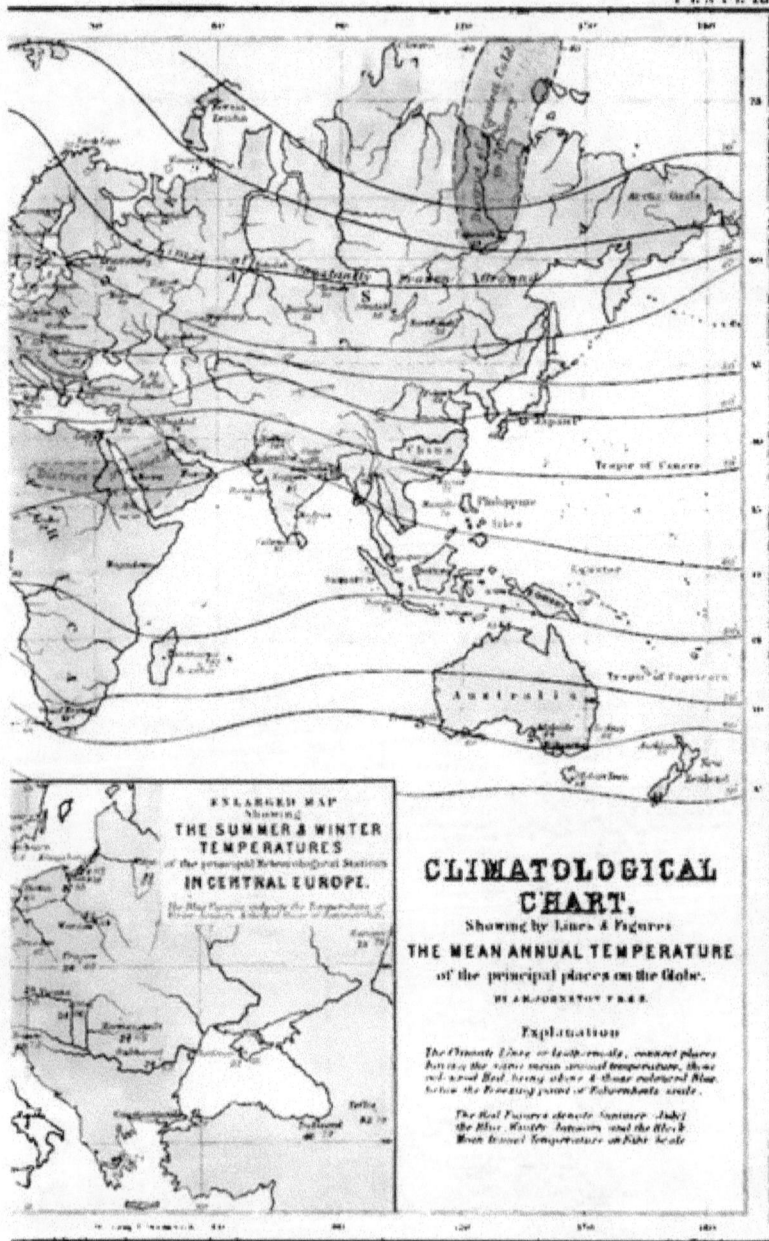

ENLARGED MAP
Showing
THE SUMMER & WINTER
TEMPERATURES
of the principal Meteorological Stations
IN CENTRAL EUROPE.

The Blue Figures indicate the Temperature of
those Januarys & the Red those of Summertime.

CLIMATOLOGICAL
CHART,
Showing by Lines & Figures
THE MEAN ANNUAL TEMPERATURE
of the principal places on the Globe.
BY A.K.JOHNSTON F.R.S.E.

Explanation

The Climate Lines or Isothermals, connect places
having the same mean annual temperature, those
coloured Red being above & those coloured Blue
below the freezing point of Fahrenheits scale.

The Red Figures denote Summer July,
the Blue Winter January, and the Black
Mean Annual Temperature on Fahr Scale.

PLATE 17

AUSTRALIA

N. POLAR WINDS

DICAL & VARIABLE

THE GLOBE

HURRICANES

CHART OF THE
RODRIGUEZ HURRICANE
In April 1843

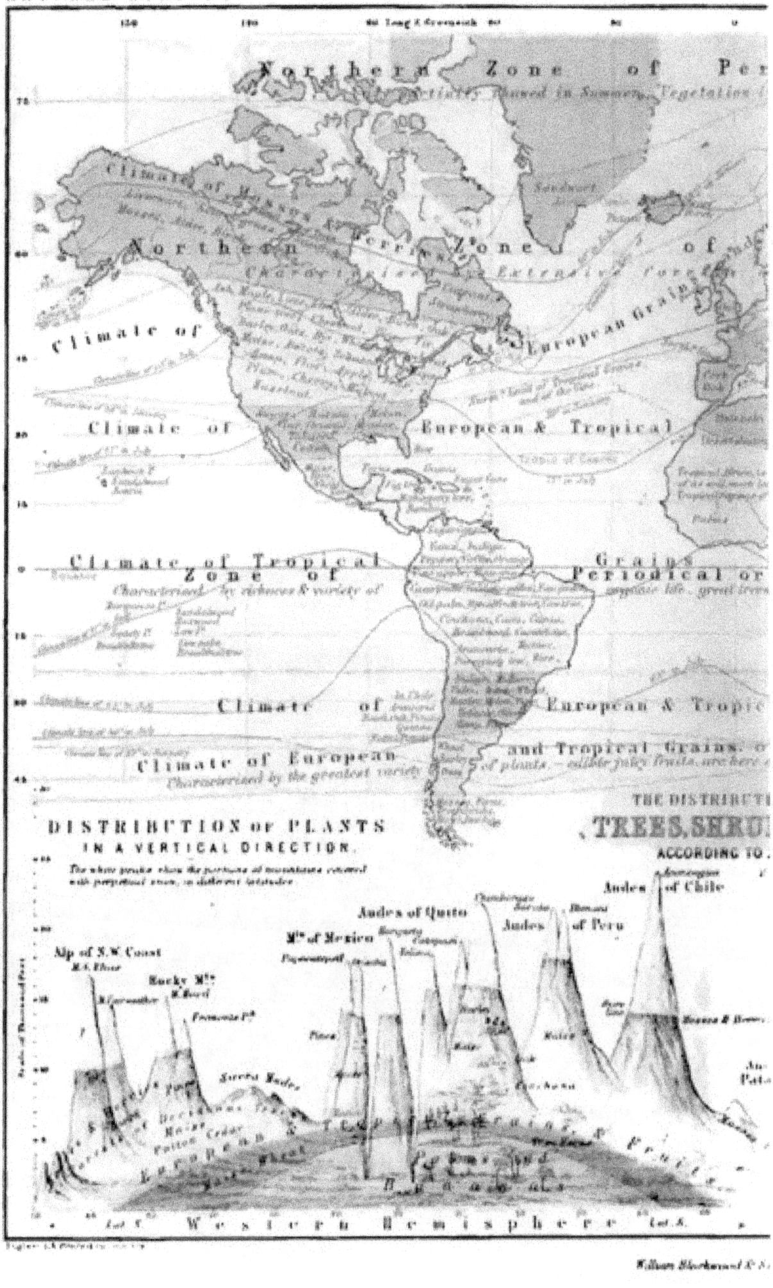
Northern Zone of Per
Climate of
Northern Zone of
European Grain
Climate of European & Tropical
Climate of Tropical Grains
Zone of Periodical or
Characterised by richness & variety of
Climate of European & Tropic
and Tropical Grains,
Climate of European
Characterised by the greatest variety
DISTRIBUTION OF PLANTS
IN A VERTICAL DIRECTION.
THE DISTRIBUTI
TREES, SHRU
ACCORDING TO
Andes of Quito
Andes of Chile
Andes of Peru
Mts of Mexico
Alp of N.W. Coast
Rocky Mts
Western Hemisphere

William Blackwood & S.

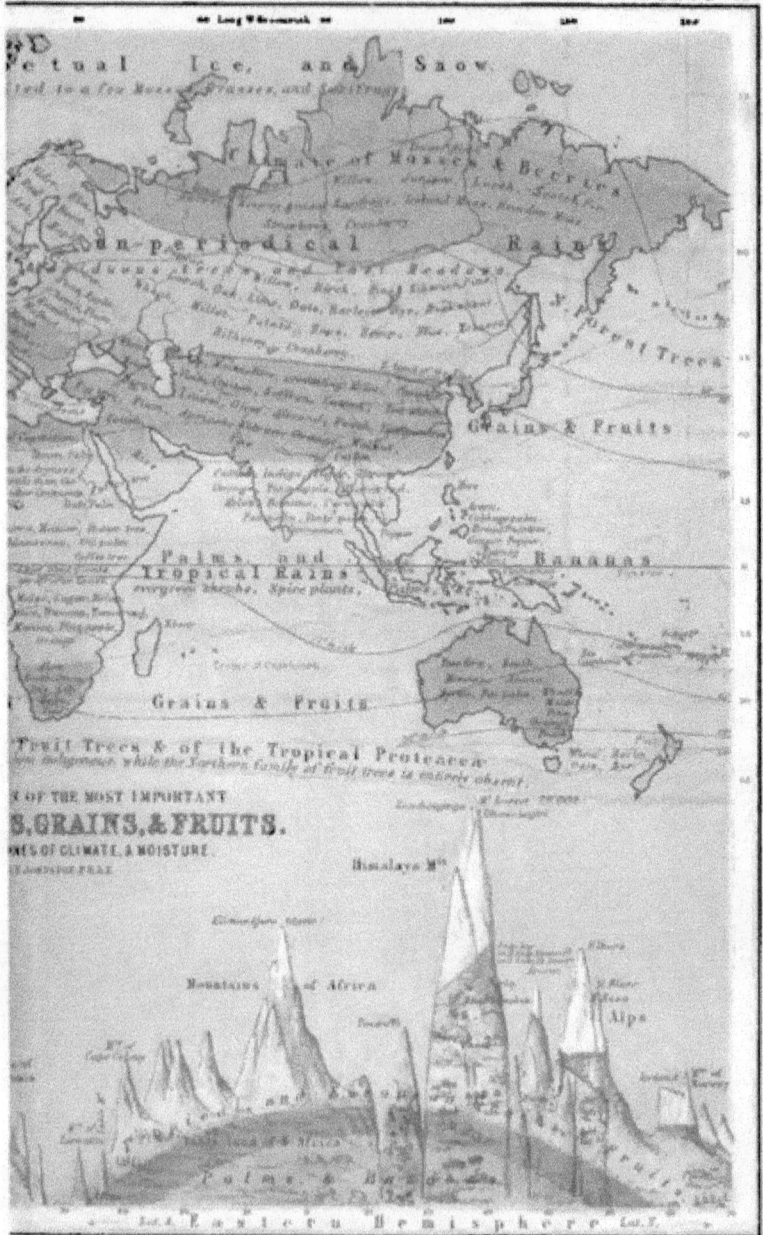

PLATE 18

Perpetual Ice and Snow

Climate of Mosses & Berries

Un-periodical Rains

Grains & Fruits

Palms and Bananas
Tropical Rains
evergreen shrubs, Spice plants,

Grains & Fruits

Fruit Trees & of the Tropical Proteaceæ

OF THE MOST IMPORTANT

B, GRAINS, & FRUITS.

ES OF CLIMATE, & MOISTURE.

Himalaya Mts

Mountains of Africa

Alps

Eastern Hemisphere

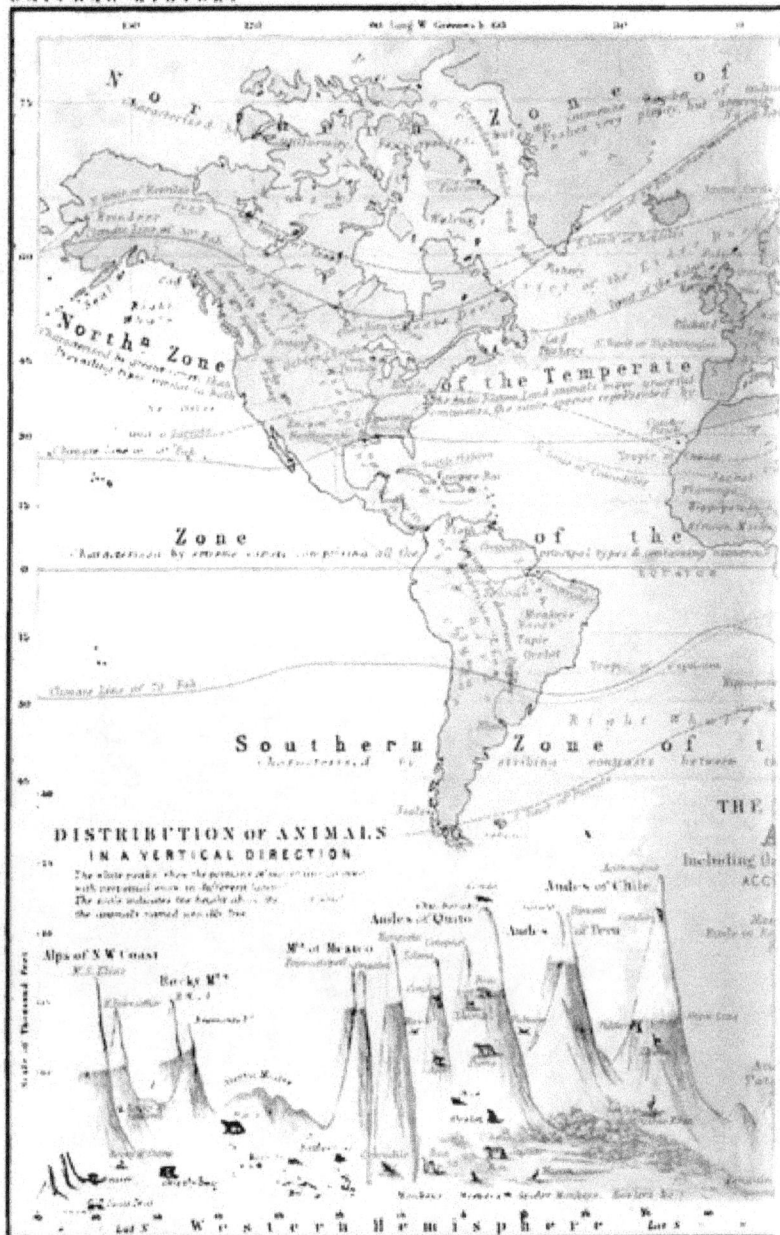

140°      120°      6th Long W. Greenwich 60°      30°

75

N    o    r    t    h         Z    o    n    e         o    f

60

45    North Zone

30

15

Zone         of         the

0                                                                    EQUATOR

15

30    Climate Line of 70 Fah.

45    Southern    Zone    of    t

DISTRIBUTION OF ANIMALS
IN A VERTICAL DIRECTION

Alps of N.W. Coast        M⁰ of Mexico        Andes of Quito        Andes of Chile
Rocky M⁰ˢ                                                          Andes of Peru

W    e    s    t    e    r    n         H    e    m    i    s    p    h    e    r    e

William Blackwood & S.

PLATE 19.

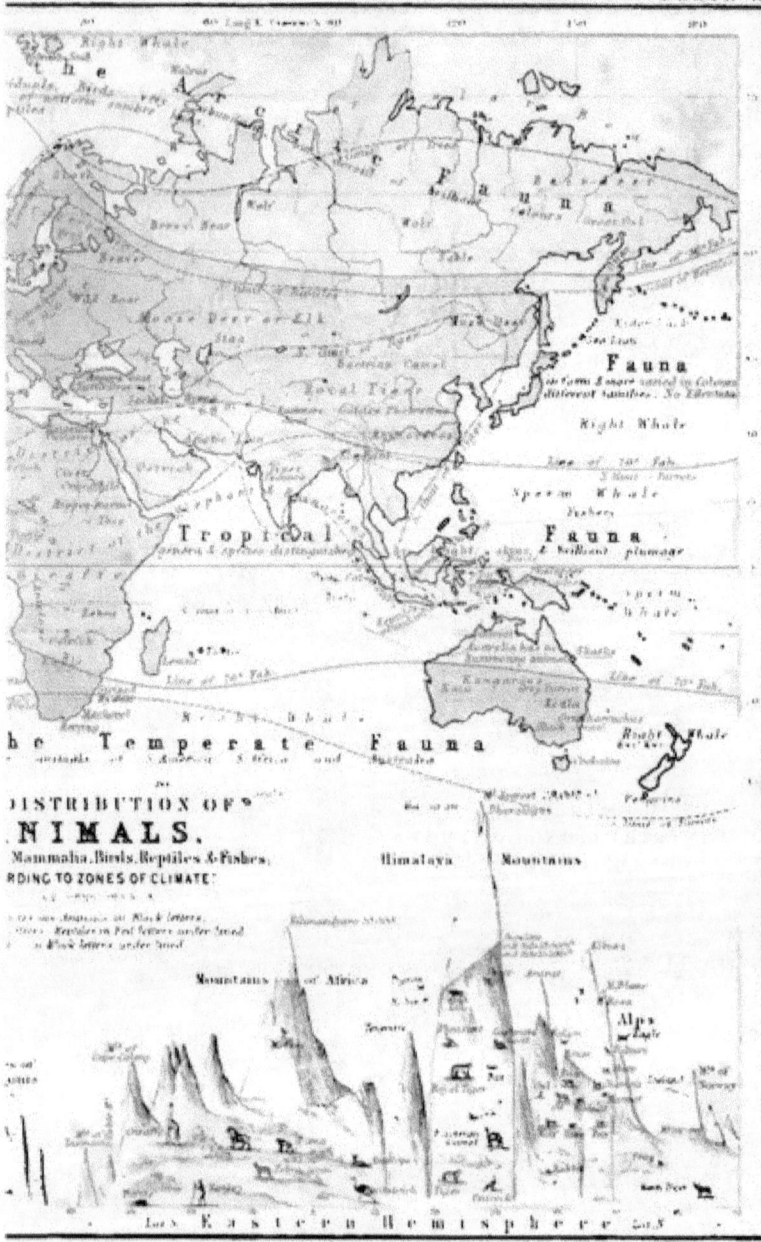

DISTRIBUTION OF
NIMALS.
Mammals, Birds, Reptiles & Fishes,
RDING TO ZONES OF CLIMATE.

Himalaya Mountains

Eastern Hemisphere

Edinburgh & London.

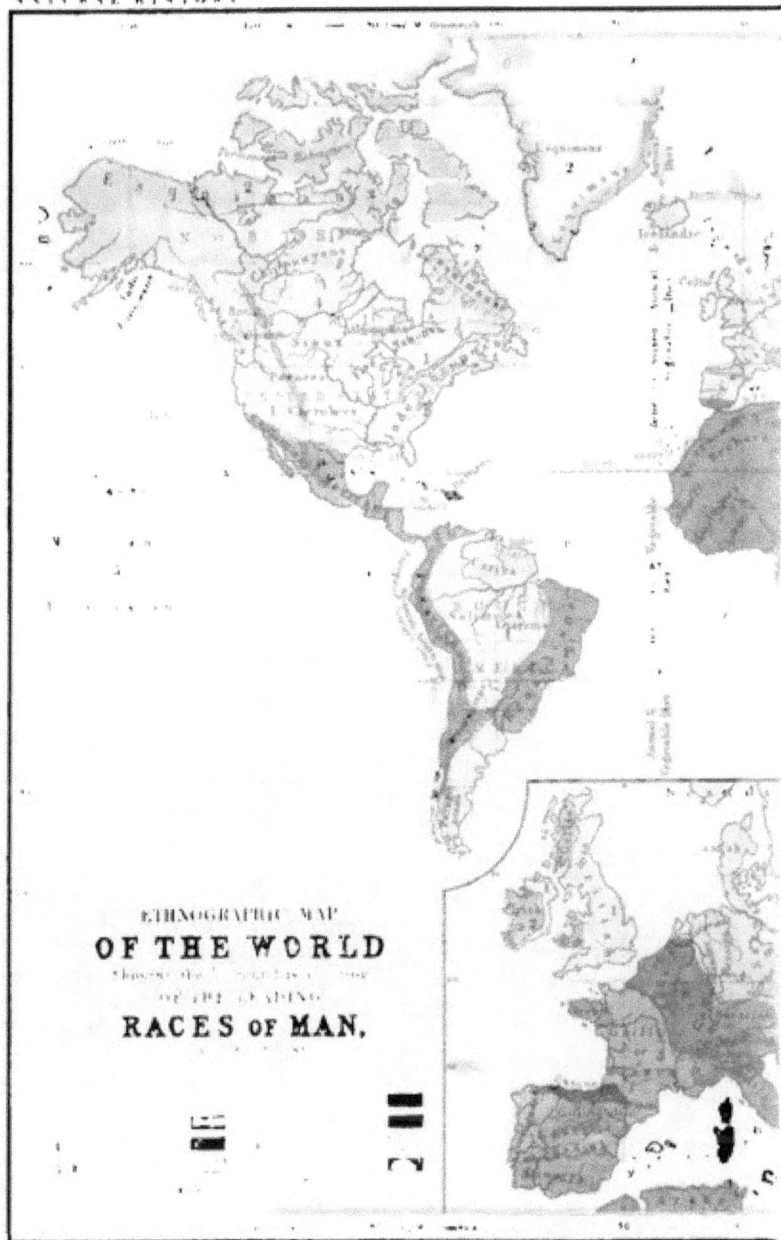

ETHNOGRAPHIC MAP

# OF THE WORLD

## RACES OF MAN.

William Blackwood & Son

PLATE 20

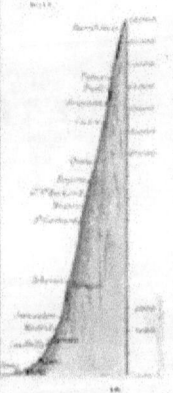

EUROPE

Edinburgh & London

www.ingramcontent.com/pod-product-compliance
Lightning Source LLC
Chambersburg PA
CBHW021821190326
41518CB00007B/690